THE PRICE
OF
FOLLY

British Blunders in the
War of American Independence

Also from Brassey's:

DAVID
Churchill's Sacrifice of the Highland Division

JACKSON
The Pomp of Yesterday: India and the Defence of Suez 1798–1918

NAPIER
Barbarian Eye: Lord Napier in China 1834 – The Prelude to Hong Kong

DIETZ
The British in the Mediterranean

LAFFIN
Brassey's Battles (New Edition)

TENNEY
My Hitch in Hell: the Bataan Death March

GILMORE
A Connecticut Yankee in the 8th Gurkha Rifles: A Burma Memoir

THE PRICE
OF
FOLLY

British Blunders in the
War of American Independence

by
WILLIAM SEYMOUR

With Maps and Plans by
WFN WATSON

BRASSEY'S
London · Washington

First English edition 1995

UK editorial offices: Brassey's, 33 John Street, London WC1N 2AT
UK orders: Marston Book Services, PO Box 87, Oxford OX2 0DT

North American orders: Brassey's Inc., PO Box 960,
Herndon, VA 22070

William Seymour has asserted his moral right to be identified as
the author of this work.

Library of Congress Cataloging in Publication Data
available

British Library Cataloguing in Publication Data
A catalogue record for this book is available from the British Library

ISBN 1 85753 018 7 Hardcover

Typeset by
The Word Shop, Bury, Lancashire
Printed in Great Britain by
Bookcraft (Bath) Ltd.

Contents

List of Maps

Preface and Acknowledgements

G O Trevelyan, and others, clearly showed that to cover in detail the 20 years or more that span the contents of this book requires many volumes. Nevertheless, for those people wanting a less minutely detailed account, I have sought to produce a simpler and more general, yet still comprehensive, survey of the major facts relating to the whole period with special emphasis on the many mistakes made by the British both in London and America. As a nation we are rather too inclined to denigrate ourselves, often quite unnecessarily. However, in the performance of the political and military leaders during the run-up to the War of American Independence, and its subsequent course, there is some excuse for indulging in that national pastime.

I am fully aware that the important part played by religious schism within the 13 colonies, with its serious interdenominational conflict, and the attitude of the Anglican establishment, should never be overlooked in any major review of the years leading up to the rebellion. It is, however, too big a subject for a limited work such as this, and was anyway something largely beyond the control of those who had the responsibility of trying to prevent or curtail hostilities. For readers wishing to delve more deeply into any particular aspects of the rebellion the source notes and bibliography should be helpful.

It has been said that wars break out by accident. The War of American Independence, which on paper lasted from 1775 to 1783 but which in fact turned from a civil war to an international one of global proportions from 1778 to 1783, did not break out by accident. Indeed it did not break out, it gradually developed as a consequence of many years of obstinacy and stupidity. There were, perhaps, even more mistakes, muddles and mismanagement than are immediately apparent during the troubled negotiations before an outbreak of hostilities and throughout the conduct of the war itself, and it is not too difficult for the historian to pass judgement in the certain knowledge of what subsequently occurred. A more difficult task, although an intriguing one, is to try to understand the problems as they presented themselves to the politicians and military commanders at the time, and to avoid the trap of judging the merits or defects of any action with the advantage of hindsight, or of evaluating the characters of those who took the actions by the standards of our own time – and this I have sought to do.

The land fighting in the war can be conveniently divided into five main phases (for details see Chronology): the opening phase of Bunker Hill, Siege of Boston and the American advance into Canada (1775–76); the fighting up what is now New York state and across the Hudson into New Jersey (1776–77); the British advance from Canada, down the Hudson to Albany (1777); the campaign in Pennsylvania and the British capture of Philadelphia (1777); and Cornwallis's Southern Campaigns of 1780–81 culminating in his surrender at Yorktown.

The chief means of making an accurate assessment of events is through the letters, diaries and documents of those with direct experience of them; but these, and individual reports also, are not always truthful, for the writer may have an axe to grind (as shown, for example, in some of Hood's correspondence), and facts can be seriously distorted or omissions made to give a more favourable impression. This is particularly so when it comes to military operations. Although this is not primarily a 'battles book', on those occasions where strategic or tactical mistakes were made, I have tried to discover the commanders' true intentions and to supply sufficient detail to make the actions clear. In writing about events, political or military, of many years ago there will always be facets that are difficult to summon from the shades, and the whole truth is never likely to be obtained. In the end, every individual has to make a personal interpretation from the positive facts that are available.

This complex, unique and wholly avoidable war, so full of blunders, mistakes, muddles and mismanagement, chiefly on the part of the British politicians responsible for its conduct, has long intrigued me – not least because General John Burgoyne was the father of my great, great grandfather, Field Marshal John Fox Burgoyne. At Appendix 1 I have included an anecdote of John Fox's own experiences in America during the futile and destructive war of 1812–15 whilst serving as a Colonel in the Royal Engineers under General Sir Edward Pakenham.

I am well aware that there have been numerous excellent accounts of this war, but so great is the interest shown by many people that I believe there is scope for this further study, concentrating on one particular aspect. There is always the temptation to ponder on what might have been if, say, the principal Opposition speakers had been in charge of affairs and (as would seem probable) rebellion had been avoided, or at worst quickly suppressed; but this would involve entering the realms of psychology and speculation, and that is a door through which I decline to go.

I am deeply indebted to a number of people who have helped me in the course of my work. Foremost among them is William Watson who has not only prepared the maps and plans without which the book would be greatly diminished, but has also read the manuscript and offered a number of invaluable sugestions as well as making some necessary corrections. I have had a great deal of assistance from the staffs of the London, British and Ministry of Defence Libraries, and the Public Record Office. In particular I would like to thank Mr Colin Stevenson of the London Library, Mrs Jean Clayton of the MoD Library, and Mr John C Dann and Mr Christopher D Fox of the William L Clements Library in Ann Arbor, Michigan. The County Archivists of Devon and Hertfordshire (Mrs Rowe and Miss Burg respectively) were also very helpful to me, as was my editor at Brassey's, Bryan Watkins, who in the course of his work made some wise suggestions and alterations that have produced a felicitous clarity to the text. Colonel Watson obtained considerable help for his work on the maps and plans from the staff of the National Army Museum, the Royal Armouries of the Tower and the National Maritime Museum.

In this field, Mr James Howard Jr of the United States National Park Service provided me with useful particulars of various battlefields, and his Park Superintendents gave much assistance at the time of my visit to the sites. The curator of the Camden Archives and Museum kindly supplied me with various details concerning the Camden battle. As always, and perhaps more than usual, I owe very much to my hard working secretary, Pat Carter, who has retyped the manuscript and source notes more times than she would care to remember. Last, but by no means least, do I thank my wife, Mary, who makes all things possible.

Dramatis Personae

British

George III, King of England

Politicians

Barré, Colonel Isaac: Opposition spokesman

Barrington, William, 2nd Viscount: Secretary-at-War 1775

Burke, Edmund: Opposition spokesman

Bute, Earl of: First Lord of Treasury 1762–63

Campbell, Lord William: Governor of South Carolina

Conway, Field Marshal Henry Seymour: Opposes government 1782

Dartmouth, William Legge, 2nd Earl of: Secretary of State 1775

Dundas, Henry: Lord Advocate of Scotland

Dunmore, John Murray, Earl of: Governor of Virginia

Ellis, Wellbore: Secretary of State 1782

Fox, Charles James: Opposition spokesman; Secretary of State 1782

Germain, Lord George: Secretary of State 1775–82

Grenville, George: Prime Minister 1763–65

Hutchinson, Thomas: Governor of Massachusetts

Lowther, Sir James: Opposition spokesman

Martin, Josiah: Governor of North Carolina 1771

North, Frederick, Lord: Prime Minister 1770–82

Pitt, William, 1st Earl of Chatham: Elder statesman

Pitt, William, second son of Chatham: Chancellor of Exchequer 1782; Prime Minister 1783

Rockingham, 2nd Marquess of: Prime Minister 1765–66 and 1782

Shelburne, William Petty, 2nd Earl of: Secretary of State; Prime Minister 1782

Townshend, Charles: Chancellor of the Exchequer 1766

Tryon, William: Governor of North Carolina 1765–71, and of New York 1771–78

Commanders-in-Chief and Senior Subordinate Commanders

Affleck, Commodore Sir Edmund: Commanded *Bedford* at the Saints

André, John: PA to Clinton

Arbuthnot, Vice-Admiral Marriot: C-in-C North American Station 1778–81

Barrington, Admiral Samuel: C-in-C Leeward Islands 1778–79

Brant, Joseph: Christian Iroquois war-chief with Burgoyne

Burgoyne, General John: Commander Canada Army 1777

Byron, Vice-Admiral John: North American Station and Leeward Islands 1778–79

Campbell, Lieutenant Colonel Archibald: Took Savannah 1778

Campbell, Brigadier General John: Commander of small force in West Florida

Carleton, General Sir Guy: C-in-C British troops Canada 1775–77

Clinton, General Sir Henry: C-in-C British troops in North America 1778–82

Cornwallis, Charles, 2nd Earl: Victor of Brandywine; 2i/c to Clinton 1778–81

Ferguson, Major Patrick: Partisan leader, killed at King's Mountain

Fraser, Brigadier General Simon: With Burgoyne at Saratoga

Gage, General Thomas: Governor of Massachusetts and C-in-C British forces 1775

Grant, Major General James: Under Howe 1776; captured St Lucia 1778

Graves, Vice-Admiral Samuel: C-in-C North American station 1776–78

Graves, Admiral Thomas (cousin to above): Succeeded Arbuthnot in command North American station July 1781

Hood, Rear Admiral Samuel: Squadron commander under Rodney 1780–82

Howe, Admiral Richard, Viscount: C-in-C North American station 1776–78

Howe, General Sir William: C-in-C British troops North America 1775–78

Knyphausen, General Wilhelm Freiher von: Commander Hessian troops

Leslie, Major General Alexander: Fought with distinction under Cornwallis in South Carolina 1780–81

O'Hara, Brigadier General Charles: Commander of Guards Regiment

Parker, Commodore (later Vice-Admiral) Sir Peter: Commander first attack on Charleston 1776; C-in-C Jamaica 1777–82

Percy, Brigadier General Hugh, Earl: Prominent at Lexington and Bunker Hill

Prevost, Major General Augustine: Commander troops Florida, Georgia; captured Savannah and Augusta 1779

Rawdon, Lieutenant Colonel Francis, Lord: C-in-C troops South Carolina 1781

Riedesel, General Adolf Friederich, Baron von: 2i/c Hessian troops

Rodney, Admiral Sir George Bt (later Lord): C-in-C Leeward Islands 1779–82

St Leger, Lieutenant Colonel Barry: Commander of Burgoyne's supporting force on eastern flank

Simcoe, Lieutenant Colonel John: Commander, Queen's Rangers

Tarleton, Lieutenant Colonel Banestre: Commander Loyalist and British Irregulars

Americans and Allies
Politicans

Adams, John: Massachusetts delegate to first Continental Congress; peace negotiator; second President of US

Adams, Samuel: Revolutionary statesman; avoided arrest at Lexington 1775

Florida Blanca, Count: Spanish Chief Minister

Franklin, Benjamin: Statesman and diplomat; principal American peace negotiator 1782–83

Hancock, John: Politician and merchant; President Continental Congress 1775–77; signed Declaration of Independence

Henry, Patrick: Revolutionary statesman; orator; elected governor of Virginia June 1776

Jay, John: Lawyer; joint peace negotiator with Franklin 1782–83

Jefferson, Thomas: Drafted the Declaration of Independence; later third President United States

Lee, Richard Henry: Moves resolutions on behalf of independence, Congress June 1776

Paine, Thomas: Political pamphleteer; author of *Common Sense* January 1776

Penn, Richard: Bearer of Congress's final address to the King 1775

Turgot, Baron: French Minister of Finance

Vergennes, Charles Gravier, Comte de: French Foreign Minister; peace negotiator 1783

Warren, Dr Joseph: President of Massachusetts Provincial Congress; killed at Bunker Hill

General Officers and Admirals

Alexander, Brigadier General William, Earl of Stirling: Prominent at Brooklyn Heights

Arnold, Brigadier General Benedict: American soldier and traitor; British general

Bougainville, Rear Admiral Louis-Antoine de: Squadron commander, Battle of the Saints

Bouillé, General François, Marquis de: C-in-C West Indies (French possession) 1778–82

Chastellux, Major General François, Marquis de: French general, served under Washington New York 1781

Conway, Major General Thomas: Inspector-General of the army December 1777; intrigues against Washington

Estaing, Charles Hector, Comte d': French *chef d'escardre* operating off North America and West Indies 1778–82

Gates, General Horatio: Army commander 1776–80; victor of Saratoga 1777

Grasse, Lieutenant General François-Joseph, Comte de: Commander naval forces North America 1781–82

Greene, Major General Nathanael: Commander southern army 1780–81

Guichen, Lieutenant General Comte de: Commander naval forces West Indies 1780–82

Howe, Major General Robert: Defeated by Provost at Savannah 1779

Lafayette, Gilbert Motier, Marquis de: Opposed to Cornwallis in Virginia 1781

Lee, Major General Charles: Experienced commander, but unreliable maverick

Lincoln, Major General Benjamin: Commander of the Southern Department 1777–80

Marion, Lieutenant Colonel Francis: Accomplished guerrilla fighter and leader of irregular troops

Montgomery, Brigadier General Richard: Former British officer; 2i/c to Schuyler in attack on Canada; killed at Quebec

Morgan, Brigadier General Daniel: Distinguished commander Saratoga and Carolina

Moultrie, Colonel William: Successful defender of Sullivan's Island 1776

Poor, Brigadier General Enoch: Brigade commander Saratoga

Prescott, Colonel William: Commanded troops (under Putnam) at Bunker Hill

Putnam, Major General Israel: Commander of troops Bunker Hill, Brooklyn Heights 1775, 1776

Rochambeau, Lieutenant General Jean-Baptiste, Comte de: Commander French troops North America 1780–81

St Clair, Major General Arthur: Defender of Ticonderoga 1777

Schuyler, Major General Philip: Commander Northern Department 1775–77

Stark, Brigadier General John: Victor of Bennington 1777

Steuben, Frederic, Baron: Inspector-General American army 1778; trained and re-modelled army; fought at Yorktown

Stevens, Brigadier General Edward: Commanded Virginian brigade with distinction at Camden and Guilford Courthouse

Sullivan, Major General John: Captured on Long Island; exchanged, and fought at Trenton, Brandywine and Rhode Island

Sumter, Brigadier General Thomas: Accomplished guerrilla fighter and leader of irregular troops

Thomas, Major General John: Captured Dorchester Heights 1776; Commander troops in Canada May 1776

Vaudreuil, Vice-Admiral de: Squadron commander, Battle of the Saints

Washington, George: Commander-in-chief throughout the war

Washington, Lieutenant Colonel William: Distinguished cavalry commander

Wayne, Brigadier General Anthony: 'Mad Anthony'; the most ubiquitous fighting general of the war.

Subordinate commanders and individuals

Allan, Ethan: Together with Arnold attacks and takes Ticonderoga 1775

Cadwalader, Colonel J: Prominent at Trenton

Dearborn, Henry: Commanded Light Infantry at Freeman's Farm

Gansevoort, Colonel Peter: Defender of Fort Stanwix 1777

Gridley, Colonel Richard: Responsible for defence layout Bunker Hill

Herkimer, Colonel Nicholas: Killed leading Fort Stanwix relief force

Kosciuszko, Thaddeus, Brigadier General: Organiser of Saratoga defences

Lee, Colonel Henry: Commander of light troops under Greene in Carolina

Warner, Seth: Commander Green Mountain Boys

Wilkinson, Lieutenant Colonel James: Deputy-Adjutant-General to Gates at Saratoga

Chronology

1650–1774

(Items in italics indicate events occurring in England or elsewhere outside America.)

1650–63   *Passage of three Trade and Navigation Acts to control smuggling and protect British and colonial ship owners against foreign competition.*

1696   *Foundation of the Board of Trade and Plantations.*

1756–63   The French – Indian Wars.

1760   *Accession of George III.*

1763   *Lord Shelburne becomes President of the Board of Trade (later assuming responsibility for American affairs) and George Grenville Prime Minister and First Lord of the Treasury (his decision to increase colonial legislation and introduce taxation in America begins the glissade to war).*
Creation of the Proclamation Line dividing the colonists and Indian tribes.

1764   *Revenue (or 'Sugar') Act (Stipulating that duty should be paid to London and in coin).*

1765   *The Stamp Act (impost on all legal and business documents)* creates resentment.
The Sons of Liberty go on the rampage in Boston.
Benjamin Franklin, agent for Pennsylvania, visits London.
First General Congress meets in New York.
Grenville replaced by the Marquess of Rockingham.

1766   *Repeal of the Stamp Act. Replaced by the Declaratory Act giving Parliament the right to tax Americans on any issue.*
The Duke of Grafton replaces Rockingham and Townshend.
Takes control of American affairs, introducing the Revenue Act and establishing a Board of Customs Commissions in Boston.
Widespread dissatisfaction and unrest, particularly in Boston, leading to the despatch of troops to the city.

1770   *Lord North replaces the Duke of Grafton.*
(5 March) The Boston Massacre.
Parliament removes all duties except on tea.

1771   *Lord Sandwich becomes First Lord of the Admiralty.*
Crushing of the 'Regulators' revolt in North Carolina.

1773   (October) Introduction of the compulsory importation of tea.
The inhabitants of Philadelphia draw up their eight resolutions against the measure. Boston, New York and Charleston join in the

1773   resistance and the coercion of agents of the East India Company.
cont.   (16 December) The Boston Tea Party.
1774   *North introduces the Regulating Act.*
        (Spring) Closure of the port of Boston and removal of customs to
        Salem.
        (5 September – 26 October) 1st Continental Congress meets in
        Philadelphia.
        General Gage becomes Governor of Massachusetts, bringing four
        battalions and orders to pacify Boston and enforce the acts.
        New militia companies formed and drill and training put on a
        regular basis.

1775

Limited impressment for the Royal Navy allowed. Strength increases from 18–28,000
        men. The Army's strength now 48,647. An increase of 55,000 agreed and the
        King agrees the hiring of mercenaries from Europe for service in America.
Introduction of the Fisheries Act and Conciliatory measures.
An increase of the Boston garrison to 10,000 men agreed.
Massachusetts Provincial Congress, meeting in Cambridge, authorises the
        impressment of 13,600 militia and orders the siege of Boston.
(19 April) Lexington and Concorde.
(10 May) Capture of Ticonderoga by the 'Green Mountain Boys'.
(Late May) Reinforcements for Gage and three major generals (Howe, Clinton
        and Burgoyne) arrive in Boston Harbour.
(15 June) 2nd Continental Congress in Philadelphia authorises the raising of
        the Continental Army and select George Washington to command it in
        the rank of General.
(17 June) Battle of Bunker Hill.
(3 July) Birth of the Continental Army. Washington assumes command.
(13 August) *Richard Penn arrives in London as an emissary of Congress with a peace*
        petition for the King, who refuses to see him.
(September) Gage recalled, and replaced by William Howe.
(November) *Before handing over as Secretary of State for the Colonies to Lord George*
        Germain, Lord Dartmouth sends Commodore Sir Peter Parker to Cape Fear
        (North Carolina) with troops and a fleet of 10 ships for Clinton.
(2 November) Canada. St John's capitulates to Montgomery's Americans.
(10 November) Arnold reaches the St Lawrence and is on the Heights of
        Abraham by the 14th.
(31 December) Montgomery and Arnold decisively defeated by Carleton at
        Quebec.
(December) Clinton leaves Boston with a handful of troops for Virginia to await
        Parker.

1776

(27 February) Caswell's Patriots defeat McLeod at Widow's Creek (North Carolina).

(4 March) Washington's General Thomas siezes Dorchester Heights, making Boston untenable.

(17 March) Howe abandons Boston, finally sailing for Halifax, Nova Scotia on 27th.

(3 May) Parker, after long delays due to bad weather, arrives off Cape Fear.

(May) Clinton and Parker, urged on by the Governor of South Carolina, plan to attack Charleston.

(28 June) Failure of Clinton and Parker's attack on Charleston.

(2 July) General Howe arrives at Staten Island to prepare for the capture of New York.

(4 July) The American Continental Congress declares Independence.

(12 July) Admiral Viscount Howe, with a large fleet, joins his brother for the attack on New York.

(July/August) *Germain sends massive reinforcements to Howe.*

(August) Clinton and Cornwallis, with 2,000 men, join Howe from the South.

(22 August) Howe lands 16,000 men on Long Island.

(27 August) Howe decisively defeats Putnam in the Battle of Long Island.

(15 September) Washington having decided to abandon New York, Howe lands on Manhattan Island unopposed.

(19 September) The brothers Howe attempt to achieve a peaceful solution to the war but the concessions they offer fall well below the minimum demands of Congress.

(28 October) Howe captures Chatterton Hill.

(16 November) Knyphausen takes Fort Washington and Cornwallis then takes Fort Lee.

(30 November) Howe submits his plan for the isolation of New England to Germain.

(1 December) Clinton seizes Rhode Island unopposed, opening the way to the Hudson River.

(6 December) Howe joins Cornwallis in Brunswick and they advance to the Delaware. (Both then return to New York).

(20 December) Howe sends Germain a second plan with emphasis on Pennsylvania and the capture of Philadelphia.

(25/26 December) Washington crosses the Delaware and wins a resounding success at Trenton.

1777

(28 February) *Burgoyne submits his 'Thoughts' (on Canada) to Germain*

(*3 March*) *Germain informs Howe of the King's approval for his second plan.*

(2 April) Before receiving Germain's letter of 3 March, Howe writes to him again, announcing his intention of approaching Philadelphia by sea.

(6 May) Burgoyne arrives in Canada and assumes command of the northern army.

(8 May) *Howe's second dispatch reaches London. Germain replies on 18 May approving Howe's proposal but adding a rider to the effect that the operation must be completed in time for Howe to cooperate with 'the army ordered to procede from Canada'.*

(24 May) Howe receives substantial reinforcements from England.

(25 June) Burgoyne's army departs from its base on the Bouquet River. At about this time, St Leger sets out for the Mohawk Valley.

(30 June) Howe embarks his army for Staten Island.

(1 July) Fraser's Advances Corps contacts the enemy covering Fort Ticonderoga.

(3–5 July) Burgoyne captures Ticonderoga and St Clair, with his garrison, escapes to join General Schuyler.

(11 July) Burgoyne sets out for Fort Edward by the direct overland route.

(13 July) St Clair, having clashed with Fraser en route, joins Schuyler at Fort Edward.

(23 July) Howe leaves New York Harbour with 260 ships and 13,799 men for Delaware Bay, without waiting for Germain's approval.

(30 July) Burgoyne's main body reaches Fort Edward, which has been abandoned. Schuyler, in a defensive position north of Stillwater, is reinforced by Washington with Generals Lee and Arnold.

(12 August) Baum leaves for Bennington, West Vermont.

(15–16 August) Battle of Bennington – a severe defeat for Baum and Burgoyne.

(23 August) After several sharp encounters, St Leger is forced by Arnold to withdraw back to Canada.

(25 August) After 42 days at sea, Howe's troops disembark at Head of Elk.

(30 August) Howe, having received Germain's letter of 18 May, writes to him dismissing any thought of his going to the assistance of Burgoyne.

(August) Washington replaces Schuyler with Gates.

(9 September) Howe finds Washington at Brandywine Creek and defeats him there on 11 September.

(13 September) Burgoyne, still heading for Albany, crosses the Hudson.

(16 September) Howe moves towards Philadelphia, sending General Gray ahead to clear the American rearguard near Paoli.

(19 September) Battle of Freeman's Farm – a costly victory for Burgoyne.

(Late September) Washington defeated by Howe at Germantown.

1777 *continued*

(3 October) Clinton, advancing up the Hudson takes Forts Montgomery and Clinton but halts there as boatmen refuse to go further.

(7 October) Battle of Bemis Heights. Burgoyne's army hard hit, weary and short of rations.

(8 October) Burgoyne decides to fall back to Saratoga.

(12 October) Burgoyne's surrender.

(16 October) Convention of Saratoga.

(22 October) Howe's Hessians defeated at Red Bank.

(Early November) Howe bombards Mud Island until 10th.

(12 November) Cornwallis retakes Red Bank.

(October) Howe tends his resignation to Germain.

(4–5 December) Washington encamped at Whitemarsh, 10–12 miles north of Philadelphia. Abercrombie and Cornwallis both defeat bodies of Americans forward of Washington's main positions in sharp encounters.

(11 December) *Burgoyne's Saratoga Dispatch reaches London.*

1778

(January/February) *Heated debates rage in Parliament over the futility of sending any more troops to America. A modified plan for future conduct of the war, with the emphasis upon naval blockade and the safeguarding of the West Indies possessions, favoured by the King, is adopted.*

(4 February) *Germain writes to Howe confirming the King's acceptance of his resignation.*

(6 February) *The French sign a Treaty of Amity and Commerce and a defensive Treaty of Alliance with America.*

(11 February) *North tells Parliament of a proposed Plan of Conciliation and a bill is passed authorising the dispatch of three Commissioners to negotiate a peace.*

(13 March) *The French formally inform the British Government of the treaties signed with America. Parliament deems this to be tantamount to a declaration of war.*

(15 March) *Comte d'Estaing's French squadron of warships leaves Toulon for American waters.*

(17 March) Admiral Rodney assumes command of the West Indies Station.

(22 April) The Peace Commissioners, headed by the Earl of Carlisle, arrive in Philadelphia.

(8 May) Clinton takes over from Howe as Commander-in-Chief.

(Early May) Two contradictory letters for Clinton arrive from Germain, dated 8 and 21 March, containing instructions from the King. He is ordered to evacuate Philadelphia and withdraw to New York. Meanwhile, offensive operations are to be mounted against the New England coastline. The promised reinforcement of 10–12,000 men drastically reduced.

1778 *continued*

(6 June) Peace Commissioners start work. A story of gross ineptitude ensues and they return to England in November – a total failure.

(9 June) Admiral Byron leaves Plymouth to relieve Lord Howe as Naval Commander-in-Chief.

(18 June) Evacuation of Philadelphia by Clinton. Washington enters the city that evening.

(28 June) Clinton's withdrawing army meets Lee in the Battle of Monmouth Courthouse.

(5 July) D'Estaing's squadron arrives off the coast of Virginia.

(29 July) D'Estaing forces the British to destroy seven small naval ships at Newport.

(August) Admiral Byron relieves Lord Howe as Commander-in-Chief.

(9–11 August) British and French fleets involved in sporadic fighting off Rhode Island. D'Estaing withdraws to repair his damaged ships in Boston Harbour, staying there until November when he withdraws to the West Indies.

(Early November) General Grant, in command in St Lucia, repulses three attacks by d'Estaing.

(December) Washington established at Middlebrook with troops on the Lower Hudson.

(23 December) Lieutenant Colonel John Campbell lands to take Savannah and routs General Robert Howe's Patriots. Joined by General Prevost, from Florida, together they take Sunbury and Augusta and Georgia is restored to the Crown.

1779

(May) Clinton, exasperated by Lord George Germain's constant interference, writes to him in the strongest terms demanding that he should desist.

(21 June) Spain declares war.

(September) The Allied naval threat to Britain is liquidated by severe scurvy which sweeps through both the French and Spanish fleets.

(23 September) D'Estaing and Lincoln invest Savannah but are defeated by Prevost and Campbell after a hard fight.

(26 December) Clinton and Cornwallis, with 7,600 men, sail from New York to attack Charleston.

1780

(11 February) Clinton and Cornwallis land on Simmons Island.

(29 March) Clinton crosses the Ashley River.

(1 April) The first parallel is dug at Charleston.

(12 April) Banistre Tarleton's British Legion defeats Huger's Patriots at Monck's Corner.

(17 April) Battle of Martinique – Rodney's first clash with de Guichen.

(7 May) Charleston. Breaching batteries reach the third parallel and begin to bombard the town.

(11 May) Lincoln surrenders Charleston, the British taking 5,677 prisoners and 20 ships.

Clinton sends three columns to pacify the backcountry and to encourage the Loyalists to support his operations.

(15 May) Rodney's second clash with de Guichen.

(19 May) Third clash between Rodney and de Guichen forces both fleets to withdraw for repairs, followed by a lull in operations.

(8 June) Satisfied that South Carolina is pacified, Clinton sails for New York, leaving Cornwallis in charge of the Southern Campaign.

(13 August) Cornwallis joins Rawdon at Camden, having had advanced warning that a large American force under Gates is approaching from North Carolina.

(15–16 August) Battle of Camden – Gates is brilliantly defeated by Cornwallis.

(17 August) Tarleton sent in pursuit of the Continentals and defeats Sumter on the Catawber River.

(September) Rodney sails for New York, leaving Hood in command in the West Indies.

(8 September) Cornwallis leaves Camden with 2,200 men for Charlotte, leaving garrisons in Camden, Charleston and Ninety-Six.

(9 October) Battle of King's Mountain. Ferguson severely beaten by Colonel William Campbell and 1,000 Patriot backwoodsmen. At about this time, intelligence received that General Green has replaced Gates.

(9 November) Sumter defeats Major Wemyss at Fishenden Ford. Tarleton also suffers reverse.

(December) Benedict Arnold, following his aborted plot to surrender West Point to Clinton, defects to the British. Cornwallis, after a series of minor local setbacks had abandoned his march into North Carolina but the arrival of General Leslie in Charleston causes him to plan a further operation. He builds up his resources at Winnsborough.

(20 December) Arnold, commanding 1,400 provincial and militia troops. sails from Portsmouth to operate against American depots at Petersburg and Richmond.

(Late December) The Dutch enter the war.

1781

(January) Leslie's troops moving north through Camden.

(17 January) Battle of Cowpens. Tarleton worsted by Brigadier General Morgan – a very severe blow to Cornwallis's campaign.

(27 January) Rodney learns of the Dutch involvement in the war and at once seizes the island of St. Eustatius, taking enormous quantities of booty.

(1 February) Cornwallis clashes with Davidson on the Catawba River and is then held up on the Yadkin at Trading Ford. Frustrated, as Greene withdraws across the Dan into Virginia, Cornwallis returns to Hillsborough.

(15 March) Battle of Guilford Courthouse. A victory for Cornwallis over Greene but at a terrible cost in soldiers' lives.

(21 March) Sandwich warns Rodney of a French fleet mustering at Brest destined for the Americas. Rodney fails to react and de Grasse arrives in the West Indies in early May, immediately attacking St Lucia, where he is rebuffed by a gallant defence.

(27 March) De Grasse makes an unopposed landing on Tobago.

(7 April) Cornwallis, his army exhausted, hungry and sadly depleted, enters Wilmington. Three days later, he begins to march north again on his own authority, heading for Virginia.

(25 April) Battle of Hobkirk's Hill. Rawdon scores a brilliant victory over Greene.

(9 May) Camden, surrounded by Greene's troops, surrenders. Florida falls to the Spanish.

(13 May) Death of General Phillips at Petersburg.

(20 May) Cornwallis arrives at Petersburg and Arnold returns to New York.

(21 May) Start of the Siege of Ninety-Six.

(26 May) Cornwallis leaves Petersburg for Richmond in search of Lafayette. Tiring of the chase, he falls back to Williamsburg.

(7 May) Rawdon starts out to relieve Ninety-Six.

(11 and 15 June) Clinton instructs Cornwallis to form a defensive post at Williamsburg or Yorktown.

(20 June) Greene calls off the siege of Ninety-Six and, pursued by Rawdon, withdraws across the Saluda.

(30 June) After a prolonged reconnaissance, Cornwallis proposes to Clinton that his best course is to withdraw to Charleston (proposal rejected).

(4 July) Battle of Green Spring Farm. After a costly fight for both sides, Lafayette withdraws.

(23 July) A change of orders again for Cornwallis. He is ordered to prepare either Old Point Comfort or Yorktown for defence.

(August) Admiral Graves replaces Arbuthnot as naval Commander-in-Chief.

(4 August) Cornwallis begins work on Yorktown's defences.

(10 August) Admiral Hood, with 14 ships of the line, leaves the West Indies for Chesapeake Bay, arriving five days ahead of de Grasse.

1781 *continued*

(30 August) De Grasse arrives with a strong fleet and body of troops off Yorktown.

(2 September) Washington enters Philadelphia unopposed, having outwitted Clinton.

(5–9 September) The Battle of the Capes. Graves and Hood worsted by de Grasse. Graves withdraws to New York for repairs.

(8 September) Battle of Eutaw Springs (South Carolina). Greene again the loser but both armies are no longer fit to continue operations.

(9 September) French and Americans begin to bombard Yorktown.

(11 October) Cornwallis warns Clinton of the urgent need for relief by land and sea.

(15 October) Cornwallis reports to Clinton that the situation in Yorktown is critical.

(16 October) A gallant sortie by the garrison brings very temporary relief. Cornwallis attempts a break-out across the river but is defeated by the weather.

(17 October) 10.00 am. Cornwallis sends a note to Washington under a white flag requesting a cessation of hostilities.

(19 October) Surrender of Yorktown.

(25 November) *News of the surrender reaches London. Fox, Barré, Burke and the Younger Pitt, supported by Burgoyne, press for a change of ministry and an end to the war.*

(Early December) Hood returns with his fleet to the West Indies.

(12 December) *Lowther's motion in the Commons to end 'all further attempts to reduce the revolted colonies' is defeated. Admiral Kempenfelt engages a large French fleet hard but with insufficient strength to prevent its escape.*

1782

(2 January) *Germain instructs Clinton that all remaining posts and garrisons on the Atlantic coast are to be retained.*

(23 January) Hood sails with 500 men under General Prescott for the relief of St Kitts.

(25 January) Hood outmanoeuvres de Grasse brilliantly, but the need for the relief force is not accepted by the garrison commander.

(7 February) *Furious exchanges in the Commons over Sandwich's and Germain's handling of the war at sea and on land. Germain later forced to resign.*

(12 February) After a gallant fight, the garrison of St Kitts is overwhelmed.

(13 February) By another adroit manoeuvre, Hood slips away from de Grasse, rejoining Rodney off Antigua on 25 February.

(10–12 March) Rodney and Hood clash with de Grasse off the Isles des Saintes.

1782 *continued*

(20 March) *North's ministry collapses and he is replaced by Rockingham.* A substantial convoy, which Rodney has failed to intercept, reaches de Grasse at Fort Royal.

(12 April) The Battle of the Saints. Not only is de Grasse's fleet hard hit but his plans for the capture of Jamaica are confounded. British mastery of the sea in the West Indies is confirmed. *For the rest of the year, negotiations for a peace settlement drag on in Paris.*

(1 July) *Death of Rockingham, who is replaced by Shelburne.*

(October) *Franco-Spanish fiasco off Gibraltar.*

(30 November) *Provisional articles of peace between Britain and America signed in Paris (Treaty of Paris).*

1783

(20 January) *Treaty of Versailles, between Britain, France and Spain.*

(15 April) Congress ratifies the Treaty of Paris.

(25 November) Last British troops leave New York. Washington, accompanied by Governor George Clinton enters the city.

CHAPTER 1

Colonial and Political Background

THE BRITISH COULD never have won the War of American Independence. The rebellion could probably have been crushed initially by firm action under a commander-in-chief who was not governed by a policy that vacillated between conciliation and suppression and which imposed upon him the dual role of commanding general and peace commissioner. With the rebellion crushed, it might have been possible to make a political settlement short of independence whereby the Colonies could have remained within the British Empire for a further period of years of mutual benefit. But once the government had embarked upon a long war of attrition, which it tried to run from a distance of 3,000 miles with its soldiery fighting in such unusual conditions as primeval forest, fever infested marshy swamps, and with a supply system that could never be entirely relied upon, the outcome was fairly predictable. However, and more importantly, it is perfectly possible that with greater understanding and more sympathy by the governing for the governed, amicable and satisfactory negotiations could have prevented a disastrous war from ever starting.

Britain began to colonise America around 1610, and in the early years the emigrants comprised in the main radicals and dissenters who had decided to abandon their native land, lured by the promise of political and religious privileges in a new one. The Puritans and Quakers settled mostly in what became New England, while the adventurers and royalists tended to carve out estates in Virginia and Maryland. The case of Maryland is interesting. The original request to found a colony was made in 1632 and granted by Charles I to George Calvert, first Lord Baltimore, an avowed Roman Catholic who wished it to be a refuge for his co-religionists. But Calvert died the same year and the charter was granted to his son to be established from part of the Virginia grant, and to be founded as a Protestant colony. However, the ambiguous wording of the charter allowed Lord Baltimore, the colony's 'Proprietor', sufficient authority to rule a colony semi-autonomous from the crown in which English Catholics would have freedom of worship.[1]

Later colonies were established in the Carolinas, Pennsylvania and New Jersey from fresh emigrants or from New Englanders who might move south when new opportunities occurred. By the time of the Restoration, emigration from England was beginning to dry up, but the population of the English

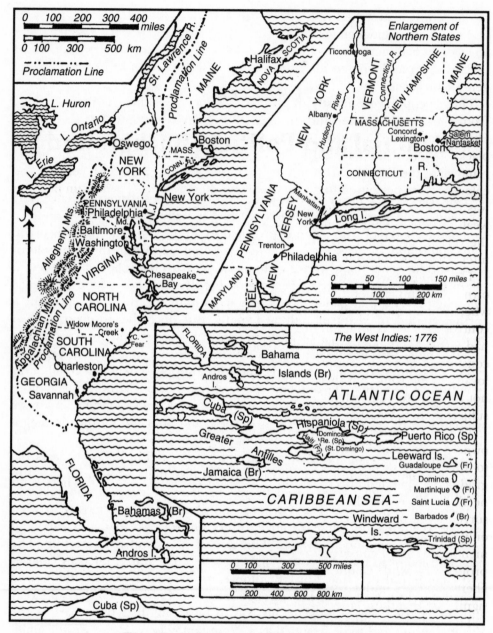

The North American Colonies, 1776

colonies, which stretched right along the Atlantic seaboard, doubled every 25 years.[2] Of an original emigration figure of perhaps 100,000 the population of 13 colonies had increased by the time of the Revolution to about two and a half million[3] men and women of native stock whose forebears had settled several generations back.

These people were mostly of British descent; the Dutch had been driven from the mouth of the Hudson and few, if any, were left, but there were Huguenots in the Carolinas, some German emigrants and about half a million blacks. Much of the later difficulties sprang from first beginnings. Up to the removal of the French threat in Canada in the middle of the 18th century, the colonists were permitted to do very much as they pleased. Their main purpose was commercial; they were an important outpost of trade, and a source of wealth to the mother country. For this reason there were restrictions, through the various navigation acts, on their trading with foreign countries, but little interference in their domestic affairs, and a minimum attempt at control.

The colonists set up their own legislative bodies and their courts, raised their own taxes, and created their own currency. It is not surprising, therefore, that when Britain developed imperialistic tendencies, and began to exert greater control over her colonies in the interests of a united empire, these people who had virtually attained independence in all but name should resent what they considered to be an erosion of their liberties. There was no question of their loyalty to the King, but this did not extend to a Parliament in which they were not represented and which wanted to order their internal affairs, and even to subject them to taxation.

The initial settlements in America were simply large feudatory estates. Their owners were granted extensive powers and privileges by royal charters, and left to their own devices to develop the land and form communities which gradually expanded into royal, proprietary or corporate colonies. There is no need to trace the history of those charters, some of which were later surrendered or annulled and revised, but by the middle of the 18th century only four of the thirteen colonies were not Crown Colonies. Maryland (already mentioned) and Pennsylvania (given by Charles II to William Penn) remained proprietary, while Connecticut and Rhode Island retained their original most liberal charters under which they elected their own governors.[4] It is perhaps significant that these latter two colonies proved to be the most loyal.

The governors, either appointed from London or locally elected, were the principal link between colony and crown. Unfortunately, with a few notable exceptions (mainly colonists of integrity and understanding) they were a poor lot. Quite often they were men whom court or family were anxious to see in employment out of Britain; they were loyal, but often capricious and tyrannical, and usually out of sympathy with the people they governed. Until changed by Parliament, their pay and consequently their behaviour, was controlled by their colony's Assembly. As a result they were frequently in contention with that body. Nevertheless, their reports to the Lords of the

Committee of Trade and Plantations (later The Board of Trade) were the chief source of information available to the government, and should have been more widely used.

The home government was inclined to regard colonial Assemblies rather in the nature of county or borough councils, whereas the colonists considered them more akin to Parliament for, as already noted, their upper and lower houses had, between them, fully comprehensive powers of government. Even a most able governor found it nearly impossible to steer a peaceful course between the expectations of his royal master and the jealously guarded charter rights of an Assembly. Governors of lesser intelligence and lack of vision were soon reduced to furious fulminations, and (often) a losing confrontation. If matters got really bad there was the final resort of the royal veto. This was a restriction that the colonists were never able to remove, but one that it would have been impractical to use.

To understand the mistakes that were made by successive British governments in their relationship to the colonies it is necessary to appreciate the difficulties that had to be faced. It is never easy to throw a light on personalities and related problems that glimmer faintly in the far shadows, but there are many windows that illumine the 17th and 18th centuries. One of the greatest difficulties was the distance that separated the mother country from her colonies. Three thousand miles of turbulent ocean in those days might be crossed – east to west – in about six weeks if conditions were favourable, but two months was more usual, and often longer was needed. The passage west to east was easier, and could be done in a month or a little over. Obviously, the long delay in the turn-round for dispatches created some misunderstandings, but many of these existed outside the time factor, and in any case when the Howe brothers were in command on both the land and the sea, during the war, instructions were apt to be disregarded. It was not only the time that ships took, but the conditions prevailing. Sickness and death took heavy toll of both men and horses in transit, particularly the latter.

Then there was the difference in the two peoples. Basically they were of the same stock, but the post-Restoration Englishman was a different person from the Puritan who had left his shores earlier, and whose New England descendants had remained steadfast in their religion and ideals. (New England held the primacy among the colonies, and Massachusetts held the primacy of New England.) There was little, if any, meeting of minds between the two peoples. The British masses, if they gave the matter any thought at all, may have regarded the distant land as a convenient dumping ground for the convict element among them; the landed gentry were inclined to consider the colonists as tenants in an exaggerated landlord and tenant relationship, while those who governed concentrated for the first hundred years or so solely on commercial values. Later, when policy impinged, the colonies took shape in many minds as an irritating, almost insoluble political complication that would not go away – not unlike the Irish difficulties of more recent times.

The Americans on the other hand felt that for an almost independent nation there was too much interference from the mother country. They could not forget that they had received their charters from the sovereign, and that their allegiance was to him and not to Parliament. This worked well in the early years, for the Stuart kings had more to contend with than American problems, and were happy to leave the colonies to develop their laws unhindered. But a degree of interference in the interests of trade was always necessary; unfortunately, when Parliament assumed control this interference was neither minimal nor firmly applied. However, so long as the French threat from neighbouring Canada remained, the colonists continued to receive velvet glove treatment, for governments always feared a possible pro-French rising. With a better understanding of local feeling, such fears need not have been seriously entertained, for the New Englanders had a constant dread that their French neighbours might turn upon them with a militant and proselytising Roman Catholicism. But a disruption in trade was a more realistic English anxiety.

Between 1689 and 1763 Britain fought four wars against the French, all partly on American soil. The most important of these was the French and Indian one of 1756–63 in which most of the colonies, albeit a trifle reluctantly, sent troops and many of them fought well, and incidentally gained valuable experience and training for later on. Fiske says the colonies raised, paid and equipped nearly 25,000 men,[5] but this is probably a high figure, for on the whole the military response was half-hearted, and British troops bore the brunt of the fighting. However, in war materials – especially naval requirements – their assistance was considerable. Vast quantities of timber of every kind was either used in local shipyards or sent to England, and materials such as hemp, tar and pitch were also supplied.

The empires of history were founded in many different ways and for many different reasons – power, prestige, *Lebensraum* – but *au fond* the principle motivation was always trade. And so it was, but even more so, with the American colonies, for there, unlike many parts of the world, there was no imperial need to suppress internecine war or ritual massacre, nor to combat famine and pestilence. Hence, if trade was the sole *raison d'être* for the colonies, it was at once necessary to impose a series of navigation and trade laws to ensure that the benefits of that trade should be reaped by the mother country, and were not channelled through foreign ships to foreign ports duty free. At the same time, some effort was made to frame those laws so that there were reciprocal advantages for the colonists.

In the early days, tobacco, grown principally in Virginia, was the most important commodity the colonies had to offer. Indigo and rice were also plentiful. Later, cotton, ginger and sugar were produced. It would be interesting to know from what species of tree the timber used for the supply of so many ships' masts came. The great coniferous forests were far away in the unoccupied west and north, while the likes of Loblolly, red and pitch pine were hardly mast material. The colonists were not encouraged to produce manufactured goods, for English merchants found the colonies an excellent outlet for their own wares, and later it became necessary to impose restriction on many articles, and also on the New England fisheries.

When trading first began, the kings of England had the power to impose duties on merchandise from their royal colonies without recourse to Parliament. The levy on Virginian tobacco became so heavy as to drive the colonists to seek an alternative market in Holland. Consequently, in 1621, the King ordered that no tobacco could be carried to a foreign port without first being landed in England and duty being paid.[6] As a sop to the colonies it was decreed that no tobacco was to be grown in England, and this was reinforced in the 1663 Trade and Navigation Act. Thus began intensive smuggling, which in turn led to trade and navigation acts. The first of these was passed in 1651 but the second, enacted in 1660, was the more important with its principal aim to protect British and colonial ship owners against foreign competition.[7] Most of the provisions of that act pleased the colonists, for shipping had become one of their most important concerns, but the clause that specifically forbade them to ship a wide number of commodities, known as the enumerated articles (which included tobacco, indigo and cotton), direct to European ports, was resented, for it meant a sharing of the profits after reselling instead of a colonial clean sweep.

Two other Trade and Navigation Acts were passed, in 1663 and 1673. Their articles followed much the same pattern of protecting shipping from Dutch, French and Spanish competition, and giving English merchants the monopoly of colonial commerce. Basically, the colonists did not quarrel with these early acts, but they usually contained some irritant among their provisions. For instance, in the 1663 act it was stipulated, in order to get the colonists to buy goods manufactured in England, that nothing was to be exported to the colonies from Europe except through British ports and in British ships. This they considered unjustified, and simply ignored. Nevertheless, British merchants increased their American trade in manufactured goods during the first half of the 18th century from around quarter of a million pounds to over four million.[8] This was to create a balance of payments problem for the Americans, who were importing more than they could export, and led to unsuccessful non-importation measures.

Two important events towards the end of the 17th century became the catalysts in shaping the destiny of the American colonies. By the last quarter of the century, the Dutch mercantile power had been broken. However, that of the French was in the ascendancy, and even more menacing than the Dutch, in that France was carving out an empire, part of which contained colonies adjacent to America. Her objective was to become a more potent naval and commercial power than Britain, and the French were to play a vital part along the road to American independence. The other significant event of this time was the accession of William III to the English throne, and the constitutional changes that came with him whereby much of the executive power enjoyed by the Stuarts passed to Parliament. As the latter grew in power, it left much colonial business to the Board of Trade and Plantations, founded in 1696, a competent body with knowledge of the plantations. Had the Privy Council and Parliament paid more attention to the advice of this body, much trouble might have been avoided.[9]

The French challenge, and the development of competitive markets which it brought, meant that England had to rely to a greater extent than hitherto on the raw materials produced by her colonies. It became even more necessary that the colonists should appreciate that, as part of the Empire, they were a dependency, and could not be permitted to trade solely for their own benefit. Inevitably, as they became increasingly prosperous and politically autonomous, they found this and other restrictions extremely galling. It was generally felt that merchants, and others, in England were getting rich on the backs of hard working colonists. They could conceive of no reason why their goods should be subjected to taxation in English ports, or why they were compelled to buy many of their requirements from English merchants, who, having a captive market, could demand exorbitant prices.

However, many of the trade laws and regulations could be rendered nugatory through large scale smuggling. The Molasses Act of 1733[10] was a case in point. This was an act whose principal purpose was to persuade New Englanders to buy their molasses and rum from British planters in the West Indies colonies, but it was a heavy blow to the colonists who previously had bought from the French and Spanish at a much cheaper rate. They immediately resorted to smuggling, on such a scale that within three years the act had become a dead letter.

This smuggling reached its apogee during the Seven Years' War when many of the colonists were actively contributing in arms to the British war effort in America, while others – notably from the middle states – were engaged in supplying the French colonies with vital commodities that the British naval blockade was attempting to deny them. The very large loss of revenue from smuggling, the cost of the war, and the need to find money for developing and protecting the new lands taken from the French, which were unlikely to pay their way for some time, produced a grave financial problem for the British government.

Englishmen were already paying higher taxes than Americans, and it was unlikely that landowners would stand for a still higher rate, while merchants would not welcome extra taxation for the protection and long term development of the sprawling forests and treacherous quags that they imagined lay west of the Allegheny mountains. And so it was that ministers in London turned their attention to measures that might put an end to the loss of revenue through smuggling, and to means whereby the Americans could contribute towards the cost of their own defence and administration. The year 1763 was to be a turning point in the colonies' history.

Three years earlier, George III, then 22 years old, had succeeded his grandfather on the throne. His youthful years had been entirely dominated by his mother's friend and adviser the Earl of Bute, whom he adored. Bute was of little consequence politically, but his charm and understanding undoubtedly helped the young king to overcome his initial lack of confidence, although it was not until George broke away from Bute that he began to make a positive impression.

George III was uneducated, but he had a certain natural ability and a very considerable capacity for hard work. He was an obstinate man who held rigid ideas on kingship, and from his early years was determined to rule – although through Parliament. At the time Parliament had not yet attained its full powers as envisaged in the settlement of 1688, and the king was still the chief executive, who directed the policy, appointed his ministers and looked to Parliament for consent, and occasionally dissent.[11]

Most of George III's ministers did not possess strong enough characters to oppose successfully his outrageously held views which centred mainly on his determination to uphold the authority of the British constituion in its entirety, and the integrity of the British Empire. Pitt (who with the Duke of Newcastle headed the ministry at the time of his succession) could have been of help to him, but Pitt was not an easy man to work with, was ailing, and was anyway disliked by the King.

All along, George was prejudiced against the colonists, who being a part of the empire, he considered should be obedient and submissive to the authority of the King in Parliament. He expected his ministers, who were answerable to him independently, to share his hard line towards the Americans, and to uphold his principles, which could at times be confused with prejudice. George certainly did not want war, but he must bear a major responsibility for it.

Bute, who, as adviser to the Princess of Wales, had been on the fringe of public affairs for some time, was quick in fulfilling his ambition to become First Lord of the Treasury soon after George's accession. He had neither the stomach nor the proficiency for the job, but his role as the young king's close confidant was important. When a change of ministry was decided upon in 1763 it was on Bute's recommendation that the triumvirate of Grenville, Lords Halifax and Egremont came to power. And perhaps more importantly, so far as the colonists were concerned, Lord Shelburne was made President of the Board of Trade, later taking charge of the department dealing with American affairs. With Pitt, whom he greatly admired, he was one of the few statesmen to have some knowledge of, and sympathy for, the Americans.

George Grenville, who now became First Lord of the Treasury, was an arrogant, fussy, industrious man of great intellect but little perception. During his ministry, measures were taken of a sufficiently unwise nature as to lay the foundations of rebellion. 1763 saw the beginning of an ever increasing glissade into avoidable war. The French threat had been decisively removed from the American continent, and Grenville considered the time had come to increase legislation and introduce taxation. The colonists, on the other hand, while remaining loyal to the king, were determined to resist any attempt to reduce their privileges or to impose new restrictions. In the course of the next two years or so, the home government was to introduce measures to put a ban on settlement in the recently acquired lands, to enforce the hitherto laxly applied Acts of Trade (particularly the French and Spanish West Indies trade), to restrict the colonial paper currency, and to raise money through taxation for the maintenance of a British defence force.

The Proclamation Line of October 1763 was the direct result of a serious rising under the Indian chief, Pontiac, who saw his ancestral lands endangered.[12] He had captured almost all the frontier forts, annihilating their garrisons, and the British government, anxious to preserve Indian territory, and realising that the Americans were unable to defend the frontier, decreed a dividing line between the two peoples along the watershed of the Appalachian mountains. The colonists regarded this ban on their territorial expansion as a further erosion of their liberties, and an excuse for the British to introduce troops that they did not want. The fact that there was new land available in recently acquired Florida did not mollify them.

It did not take Grenville long to discover that the casual manner in which the trade and navigation acts were being administered was costing the country some four times the amount of revenue received. This was, of course, due to smuggling. Steps were taken to increase the naval and revenue cutter presence on the American seaboard with the right to search and seize ships found to be conducting illegal trading; a change was made in the composition of the colonial vice-admiralty courts that dealt with revenue matters, and 'Writs of Assistance' could in future be issued enabling searches to be made for contraband goods in privately owned buildings.

These and other regulations were embodied in a Revenue Act of 1764, which became known as the Sugar Act because it was aimed at stopping the evasion of duty on the French and West Indies trade, which the old (1733) Molasses Act had signally failed to do. It did reduce the molasses duty from 6d to 3d a pound – but smuggling was duty free! This reduction was partly invalidated by a new levy on commodities that had hitherto been untaxed. The act also stipulated that the duty should be paid to London in coin, of which American merchants were short, and the alternative paper currency was itself subject to restraints in another act. It is perhaps not surprising that this the first of Grenville's acts caused deep resentment among the colonists.

It has been said that the trade and navigation acts, of which of course the Sugar Act was one, were a greater cause of revolt than taxation. Undoubtedly, the interference in American commerce had a very damaging effect on colonial relations, but it was taxation that sparked the fearful riots in Boston and elsewhere which were the precursors of the war. Strangely enough, the act that, more than any other, poured vials of wrath upon unsuspecting minsterial heads, was the Stamp Act. Under consideration for at least a year, it was enacted, to a well prepared reception, in 1765.

The tax, which varied from as little 3d to £10, was a straightforward impost on all legal and business documents. Grenville reckoned it would bring in £100,000 annually. This would more than cover the cost of colonial defence, then said to be in the region of £36,000, which was its primary purpose. It was to replace the current method of raising money through requests made from London to the governors and transmitted by them to the Assemblies. One of the colonists' many objections to the tax was that that method had worked adequately; but in fact the load had been unevenly spread and the revenue unreliably received, while the Stamp Tax would bear evenly on all.

Grenville had handled the preliminaries, in and out of Parliament, with patient consideration, and was anxious that the colonists should express their views after time for consultation. He was probably greatly surprised at the violence of those views, which culminated in damaging riots, the most serious of which occurred in Boston, where organisations calling themselves the Sons of Liberty (coined from a phrase used by Colonel Barré in Parliament) went on the rampage.[13] Grenville was not alone in thinking that the Stamp Tax was quite justifiable. There were dissenting voices from such as Burke, Pitt and Barré, but most of those in authority – and not least the King – felt the act to be a fair one for, as they understood it, in an imperial system all should be taxed and certainly the colonists should contribute to their defence.

The colonists made numerous representations to London, which were reinforced by the personal visit of Benjamin Franklin as agent for Pennsylvania. External taxes, such as the Sugar Act, they agreed were permissible without representation, but not internal taxes. However, in truth there was little difference between internal and external when it came to taxation; and many Englishmen felt the slogan 'no taxation without representation' was equally relevant for them with so many rotten and pocket boroughs.

An important side issue of the act was the rise to prominence of such colonists as John and Samuel Adams, James Otis, and Patrick Henry, patriots blessed with a gifted, vital and fiery rhetoric, who set the people alight with their passionate and steel-worded exhortations in defence of liberty and privilege. It was soon realised that individual agitations achieved less than co-ordinated ones and, at the instigation of Massachusetts town meetings, a General Congress assembled in New York in October 1765. It was the harbinger of the First Continental Congress that was to meet in Philadelphia nine years later and, like that assembly, had no constitutional authority to legislate or govern, but it provided a platform for resolutions both of loyalty to Britain, and displeasure at the actions of her government. In pursuit of the latter it instructed merchants to cease from the purchase of all goods from the mother country, a measure that caused a few squeals in British shipping and trading circles at the time, but was subsequently seen to have done little harm to anyone.

It is often said that Grenville blundered into an act for which there was no justification, but that is not entirely true. The purpose of the Stamp Act was sound. There could be no reason why the colonists should not pay for their own defence, which had to be by British troops, for the past troubles had shown that the colonists were not yet ready to do the job themselves, and there was still a fairly strong Spanish and French presence on the continent that represented a threat. However, the money could have been raised in other ways. Had Grenville's well intended preliminaries been developed into full and amicable consultation, which despite the difficulties of distance would have been possible, a satisfactory solution might have been reached. But King George and his ministers usually preferred dictation flavoured with vacillation, to consultation or conciliation.

In July 1765, the King could no longer tolerate Grenville, and he was replaced by a Whig ministry under the Marquess of Rockingham. Rockingham was a fairly ineffectual man, and the ministry a very weak one. Meanwhile, the serious riots in such places as Boston and New York, and the fact that those concerned were refusing to apply the Stamp Act made it obvious that only through force could it be worked. General Gage, commanding the troops in New York, was quick to point out that this would certainly lead to civil war. Accordingly, Rockingham decided to repeal the act having been influenced also by Pitt who, roused from his sickbed, forcefully asserted that expediency, if nothing else, demanded that British sovereignty over the Americans could not extend to taxation without consent.[14] The King was very much against repeal, and wished that the act should only be modified. However, Rockingham's threatened resignation proved decisive when added to fear of the return of Grenville, whose motion to retain the act had been defeated in the House. Nevertheless, repeal might not have been achieved had not the government agreed on a face-saving Declaratory Act giving it the right, in defiance of Pitt's pleading, to tax the Americans on any issue including finance. It was an unecessary act calculated to irritate should its theoretical purpose be transformed into practice, as it was a year later.

The repeal of the Stamp Act (March 1766) was greeted in both England and America with much rejoicing. In the colonies they celebrated a triumph, loyalty to the King reached a high level (for the King's opposition to the repeal had not been realised by the colonists) and for a while scarcely a ripple disturbed the political surface. At home the merchants were glad to see the act go, for they feared the effects of continued non-importation. But in Parliament there were very few who realised the true implication of this surrender to violence and contumely.

Exactly a year after he had taken office, Rockingham, and his ministry, fell. Pitt, who had refused to join with Rockingham, now sided with the King against him, and was asked to form a government.[15] He felt too ill for the task of premier and went to the Lords as Lord Privy Seal, but he selected Lords Shelburne and Camden to serve under the nominal leadership of the Duke of Grafton. The ministry was something of a gallimaufry, including as it did, against Pitt's inclination, Charles Townshend, a clever (almost too clever), witty, energetic and unscrupulous politician. As Grafton lacked ability and energy,

Townshend soon took charge. Almost at once, he destroyed the burgeoning tranquillity that followed the Stamp Act repeal with a blunder of enormous proportions. He decided to challenge the colonists' assertion that they would accept external taxation by imposing a duty on the import of such commodities as glass, paper, wine, oil, painters' colours and tea, and announced that the revenue – which was expected to be in the order of £40,000 a year – was to pay the fixed salaries of the governors and justices. A Board of Customs Commissioners for the whole country was to be set up in Boston, and measures were to be taken through 'Writs of Assistance' to ensure the collection of the revenue.[16]

It was an act of provocation, if not revanchism, which Townshend surely knew would cause immense offence, for apart from anything else it removed from the colonists the power to control their governors and officials through the manipulation of their salaries. Further resentment was caused when it was learnt that offenders under the act were to be tried in courts without juries, and in some cases shipped to England for trial. The consequences of this Revenue Act were entirely predictable, leading to serious riots especially in Boston, where the Commissioners of Customs attempted to carry out their first duty under the act by searching a sloop, appropriately named *Liberty*, owned by a prominent Boston merchant. It was not long before the Commissioners were forced to abandon their duties, and so great was the trouble that troops had to be sent to Boston from Halifax, Noval Scotia. These comprised two weak battalions and, although they were later reinforced, they were still inadequate. If force is to be used in aid of the civil power, it must be sufficiently strong to fulfil its purpose; moreover, there must be an efficient, properly functioning civil power to direct the use of force. No magistrate in Boston would dare to be seen having anything to do with the troops, who were consequently subjected to frequent degrading, and often brutal attacks by the savage mob.

Eventually, on 5 March 1770, a patrol goaded beyond endurance opened fire without orders and six of the mob were killed.[17] The so-called 'Boston Massacre' proved a wonderful opportunity for Samuel Adams and his revolutionary party to persuade, or more accurately, by threats to compel Thomas Hutchinson, the Lieutenant-Governor, to order the withdrawal of all troops from Boston, thus leaving the way open for further mischief. The situation in both America and England, however, was becoming so bad that even the King, who had warmly supported Townshend's act, was alarmed at the collapse of trade through the colonists' renewed non-importation measures. As a result, in March 1770, Parliament confirmed the Cabinet decision of the previous May to remove all the duties, save that on tea.[18]

At the beginning of 1770 there had been another change in government, and England had her third prime minister in seven years when the King persuaded Lord North to head what remained of the Duke of Grafton's minstry. On Grafton's resignation the King refused to dissolve Parliament, for he did not wish for the return either of Rockingham or Chatham, who was now partly recovered from illness. The fact is there was no proper party government at this time. In the ten years preceding the Revolution, so called Whig and Tory ministries rose and fell with alarming rapidity, their members often split into factions and divided among themselves over important issues. The country may have been nominally a democracy, but power was in the hands of a small camarilla, a ruling élite of politicians and King's friends. It was unfortunate that most of these men (and the King too) clung to long held, myopic, no-nonsense policies for colonial government. Seldom, if ever, did they listen to the moderating voices of Pitt, Fox, Barré, Camden and Burke, or even Burgoyne and Howe, the military members of Parliament who later had to extend polemical policies into practical warfare. However, at last the King, whose influence on affairs had by now become crucial, had found a man with whom he would work for the next twelve years in comparative harmony. Lord North was a good natured man who was well liked, he was not unintelligent but, unlike his King, he lacked moral courage and the strength to preside over a ministry in which each minister ran a virtually independent department answerable directly to the King, who was not above permitting, if not encouraging, individual ministers to complain of their colleagues. Nevertheless North's ministry, with its popular support over an Opposition somewhat in disarray, gave the country much needed equilibrium.

The retention of the duty on tea alone was a serious blunder on Townshend's part. It was kept as a reminder of the right to tax, but brought in no revenue, for despite every effort to curb the smuggling of tea from Holland this continued on a grand scale, while the continued imposition of the duty formed a splendid platform for patriot agitations and was to become the flashpoint of revolution. But meanwhile there were to be three years of welcome tranquillity, and the colonists were in the happy position of paying no imperial taxes. Unfortunately, instead of spending these years in mending fences, and working out a more enlightened policy towards the colonists, the government in 1773 relumed the flame of resentment with a scheme for compelling the importation of tea. In conjunction with the powerful East India Company, whose affairs were in a parlous state, a scheme was devised to enable the Company to sell the enormous backlog of tea being held in its warehouses to the colonists at a price which was lower than that being charged for the smuggled article. This was achieved by remitting three-fifths of the duty then being paid by the Company on the arrival of its tea in England. Accordingly, an Act to effect that scheme was passed in May 1773.[19] However, the first consignments of this tea did not arrive at the American ports until the autumn, thereby giving the colonists plenty of time to brood over this new government strategem. As a result, the inhabitants of Philadelphia met in the state-house on 18 October 1773 and drew up eight resolutions indicating their fury against the Act. They then requested the East India Company's agents to resign – which most of them did. This action of the Philadelphians was followed in Boston, New York and Charleston. As a result, in most ports at which tea was eventually landed, it was consigned to warehouses and left to rot. However, in Boston the officials refused to resign and there the issue was to be dramatically and decisively resolved.

Negotiations were long and acrimonious; the town meeting, (see below) ably supported by a noisy claque, was adamant in its demands that the three ships return to England with their cargoes. However, clearance from the Collector of Customs was not forthcoming and the Governor gave orders that any ship attempting to leave the harbour without a permit should be fired on. The outcome of these confabulations was the famous Boston Tea Party, when fifty Patriots, dressed as Mohawk Indians, boarded the ships and emptied 342 chests of tea into the sea.

This gross act of lawlessness, when it became known in England, prompted Fox and others to press for the repeal of the tea duty but North, with the full support of the King, declared that 'To repeal the tea duty would stamp us with timidity'. This was correct ministerial thinking, but it was seven years too late. Instead, he introduced five coercive acts, at least three of which he must have known were certain to exacerbate an already explosive situation.[20]

In the spring of 1774 the Boston Port Bill, enforced by a strong naval presence, closed Boston harbour and transferred all the customs business to the port of Salem. This may have fulfilled its alleged purpose of better management of American affairs, but the Bostonians had no difficulty in recognising it as an act of retribution. It was quickly followed by the Regulating Act, which changed the Massachusetts charter so that members of the Upper House of the legislature would in future be nominated by the Crown, and no longer elected by the Lower House. The Governor was empowered to appoint the judges, court officials and sheriffs and to fix their salaries. But the part of this act that hit hardest, and trammelled still further the colonists' independence, was the abolition of town meetings, their means of local government. A town meeting was very similar to a British urban or parish council. Delegates were elected, or chosen, to represent the outlying towns in a province from local people such as agricultural workers, shopkeepers etc. These 'town councillors' might form committees and travel to the big city, i.e. Boston in the case of Massachusetts, for a meeting at which all the town committees would assemble.

The third of these so-called 'Intolerable Acts' broadened the governors' powers for dispatching officials for trial in England. A pointless measure, well calculated to infuriate. The fourth and fifth acts were more irritating than infuriating. The fourth concerned the quartering of troops, and the fifth, the Quebec Act, not only extended the Canadian boundary but, more annoyingly to the staunchly Protestant New Englanders, it gave religious freedom to Roman Catholics. This latter act was prompted by the British government's quite unfounded fear that the French Canadians might support the colonists' embryo attempts at union, which was taking gradual shape through the initiative of Samuel Adams and his Committees of Correspondence. This was a system whereby all the towns of Massachusetts were able to act in unity, and which gathered momentum on an intercontinental basis.[21]

In the twelve months that followed the passing of the coercive acts, the situation in America rapidly deteriorated. Thomas Hutchinson, after a brief spell as governor of Massachusetts, sailed for home shortly after the Tea Party, and General Gage was ordered to Boston as his replacement.[22] He brought with him four battalions, and his instructions were to pacify the town and enforce the acts – an impossible task with the paucity of troops under his command. The troublemakers were now in great strength throughout the colony and continually flouted authority. At the same time, they were openly collecting arms and stores, drilling a large militia, and acting towards those colonists who remained loyal to the King (referred to hereafter as Loyalists) with considerable brutality.

On a national scale all the other colonies (except Georgia and Florida) had rallied behind Massachusetts in that colony's acts of defiance. In the autumn of 1774, the First Continental Congress was formed in Philadelphia.[23] The fact that it had no constitutional authority to legislate or to govern did not deter its members from voting to abrogate all the recent acts of Parliament and to issue their own administrative orders and regulations. These included the formation of a Continental Association that bound all the colonies to a damaging industrial attack on Britain through stringent non-importation and exportation measures, and a Declaration of Rights giving every provincial assembly the exclusive right in all legislation and taxation matters. The First Continental Congress may have lacked the means to put its resolutions into practice, nevertheless in its defiance to British sovereignty it was a morale booster to the colonists, and brought war a step closer.

By the beginning of 1775, Parliament's writ was in abeyance pretty well throughout the colonies, and they had become virtually ungovernable. Gage had been saying for some time that troops in considerable numbers were necessary to carry out the policy of coercion. There had been serious outbreaks of violence towards the end of 1774, not only throughout Massachusetts but also in Rhode Island, where 40 cannon were seized, and in New Hampshire, where a small fort was taken, along with a quantity of stores.[24] Early the next year, in a further attempt to rule by repression, the government introduced a bill to compel the colonies to trade only with Britain and the British West Indies. They were not to be permitted to trade with each other, nor were they allowed to use the fishing grounds off Newfoundland. Opposition attempts to modify some of the terms of the act were defeated, but greater events intervened before it became law.

The government was obviously aware that when news of the Fisheries Act, as this latest statute was called, reached the colonies the disturbances and revolutionary fervour would be redoubled. They therefore agreed to raise the Boston garrison to 10,000 men, but this in Gage's opinion was quite inadequate and he sensibly asked for 20,000 troops,[25] only to be told that such a large force could not be found and, in the government's (mistaken) opinion, was quite unnecessary. However, with the catalyst of revolution at hand, a sensible conciliatory measure was at last forthcoming. Complete freedom from imperial taxation was promised to any colony that would make a realistic contribution to defence and the maintenance of civil authority.[26] Had such a gesture been made a year earlier; it might have averted catastrophe. But events overtook it, for it had come too late.

* * *

Such is the background of events which led up to a war that nobody really wanted. The question is: could it have been avoided? The answer must be 'yes', in spite of the British political system of the 17th and 18th centuries in which patronage, jobbery and corruption were far too common. It was a system that produced only a small number of statesmen of stature, and some ministers of mediocre quality, quite unsuitable for the contentions of these difficult times, in the service of a King who was obstinate and interfering but not tyrannical. To avoid war, the sovereign and his ministers needed to take a very different approach to the colonial question from that which they adopted in the years immediately prior to the outbreak of hostilities. It did not need men of luminous intelligence, but of patience and understanding.

The problems facing 18th century statesmen are not easily understood by those of us living 200 years later, and they cannot be blamed for failing to appreciate that colonies could have self-government and still remain a part of the Empire. Nevertheless, in 1763, the year in which troubles began to mount, there was almost universal loyalty to, and sentiment for, the mother country on which to build friendship. Both nations had much to offer, and the bonds of sovereignty could have been loosened gradually through an harmonious relationship based on the enriching and humanising effect of seaborne trade.

This was not to be, and there were many reasons for it. Foremost perhaps was the complete failure by successive British statesmen to understand the extent of the American problem or the hopes and ambitions of the American people. The colonists were a proud, progressive race of men and women rapidly becoming fully competent to govern themselves, but at the same time, like most colonials, a trifle touchy and short of the important element of self-confidence. Only a handful of English sachems had shown a proper regard to the dangers and difficulties of the time, and their advice and warning went largely unheeded. Almost as bad was the lack of interest shown in the Governors' reports. A few of these men were out of touch, but most had their finger on the pulse and their reports home gave a valuable insight into the temper of their people.

Fundamentally, the difference between the two countries lay in the system, and the completely contrasting political traditions. In a way democracy in America was more advanced than it was in England. Colonial Assemblies were a great deal more representative of the people than was Parliament. Whereas the colonists felt their legislators truly represented them, Britons had not that close affiliation with those who governed them, it usually being enough that if Parliament decreed a law or a tax it should be obeyed or paid. With the exception of a few of the more enlightened ministers, there was a feeling in Westminster that what was fair enough for the British people was fair enough for Britain's colonial subjects. And even those few who demurred at taxation were opposed to any radical change of policy which would alter a colony's status of subordination, and principal purpose of profitable trade for the mother country. This was something the colonists could never understand, or if they understood, tolerate.

These and certain other differences in the two political systems engendered the blunders that eventually led to war. For instance, Grenville had not intended his Stamp Act to be tyrannical. He had an absolute belief in right and wrong, and he genuinely believed he had the right to levy money for the colonies' defence in that way. The fault lay in his being out of touch with the people he was trying to tax, and the system of government they were striving to live under. However, had he been in a position to enforce acceptance of the tax, it was still just possible he might have got away with it. The blunder lay in imposing a tax he should have known to be unacceptable (the indications were clear), and which his successor had to revoke. After that the right of Parliament to tax the colonies was virtually lost.

In contrast to Grenville, Townshend's thinking had been naive and provocative, and the failure to withdraw the tax on tea when his other duties were removed was an enormous error. Then again, most of the navigation laws were not properly thought out, and were certain to result in confrontation, for the Americans were primarily a trading nation. Once their trade with foreign countries was stopped, it was fairly certain that they would resort to smuggling, which only led to further irritating restraints.

By the time of Lord North's coercive acts, it is doubtful if a resort to arms, although not necessarily a prolonged war, could have been avoided. His attempts at conciliation were too little and too late; but even then a royal commission sent out to listen and negotiate might have achieved at least a cooling off period from which a *modus vivendi* could have been worked out whereby the colonies would have remained under the imperial umbrella for a few more years of mutual benefit.

Disturbances and disorders, even rebellions, in the American colonies, usually with a religious connotation, were almost endemic in the backcountry throughout the years leading up to 1775. But these were comparatively minor affairs capable of being sorted out by colonial governors and assemblies, although the last and most serious, that of the Regulators (1768–71), had to be crushed in battle. These backcountry people in North Carolina had been angered by government discrimination and resorted to force in order to obtain a redress of their grievances. They were defeated in the Battle of Alamance in

1771. It was in Britain that the colonists needed to be handled with a great deal more sensitivity than George III and his ministers had shown betwen 1763 and 1775. There were many opportunities for enlightened government towards the peaceful development of an Anglo-American relationship. But to take them it needed ministers who could stand up to their king, did not dither or procrastinate, were prepared to back their policies with strength and firmness and who properly understood that, on most occasions, conciliation was better than coercion. Then the greatest rebellion of them all need never have broken out. Even had it done so, greater resolution by the military commanders, properly backed by the government, could most probably have put it down in the early stages, and so avoided a major war.

CHAPTER 2

The Shooting Starts

THE STRUGGLE FOR peace was to continue right through 1775 and into 1776, but a deteriorating political situation, and some military sparring, brought full-scale war closer every month. A brief examination of the likely line-up of the opposing forces is interesting.

As is so often the case after a long war, the British army and navy had been allowed to decline. Since 1763 the navy had been on a reduced establishment of 18,000 men, the ships were mostly in bad condition, and their crews not much better, but in 1771 Lord Sandwich had become First Lord of the Admiralty for the third time, and matters began to improve. In 1775 limited impressment was allowed, and the navy's strength on paper was increased to 28,000 men.[1] In the time allowed him the First Lord did his best, in spite of Lord North's constant urgings for economy, to bring the fleet to a state of readiness at which it was capable of meeting the most formidable threat to British sea power for many years. In particular, Sandwich improved the construction, repair and maintenance work through his reorganisation and resiting of dockyards and victualling yards, and his attention to the needs of skilled and unskilled labour. He also saw to the rehabilitating and retraining of the many senior officers who, since the last war, had been leading a fairly idle existence on half-pay or unemployed.

In 1775 the British Army numbered on paper 48,647 men of which 39,294 were infantrymen, 6,869 were cavalrymen and 2,484 were gunners. There were numerous overseas garrisons, but the bulk of the Army was distributed between England (15,000), Ireland (12,000) and America (8,580). The strength of an infantry regiment (a regiment normally consisted of one battalion) was 477 men divided into ten companies of which one was a grenadier and another light infantry.[2] They were armed with the smooth-bore flintlock musket known as Brown Bess, a weapon inaccurate over a hundred yards, and unpredictable in wet weather. The light companies carried a fusil, sometimes described as a light, rifled musket, and later in the war there were a very few rifled flintlocks in action. The artillery had six and three-pounder guns, and two were usually allotted to a regiment. There was also in America a small, but efficient corps of engineers, but the transport and medical services were very sketchy. No medical corps existed, but a surgeon and his mate were nominally attached to each regiment, the nurses often being found from soldiers' wives or molls, who were permitted to be on the ration strength.[3]

Having regard to the small number of troops in Britain at the time, Lord Barrington, then Secretary-at-War, could truthfully say it was impossible to give Gage the number of men he had specifically asked for. The King had been pressing for an increase in the Army for some time, but it was not until 1775 that it was agreed to raise the British establishment to 55,000.[4] This was a figure far too small for the needs of the forthcoming war in America, and anyway not easily obtained, for a great many families had been emigrating recently to America from Scotland, the principal recruiting ground. The King therefore agreed to replace garrison troops in Gibralter and Minorca with mercenaries, and to hire others for service in America.

The Opposition used this as another scourge with which to belabour the Government, and the subsequent employment of these foreigners was deeply resented by the Americans. Moreover, there was a little difficulty in procuring them. Some rulers who were approached, notably those of Holland and Russia, returned a blunt refusal. Lord North informed the King that the letter from the Empress Catherine was 'not in so genteel a manner as I should have thought might have been expected from Her; . . . and [she] has thrown out some expressions that may be civil to a Russian Ear but certainly not to more civilised ones.'[5] However, rulers of some German states were only too willing to sell soldiers for British cash. The Landgrave of Hesse-Kassel agreed to provide 16,992 men. These included 600 cavalrymen and three corps of artillery. Others to come forward were the rulers of Brunswick (2,422), Hesse-Hanau (2,353), Anspach-Bayreuth (1,225), and Waldeck (1,152). Altogether these German princes contributed nearly 30,000 men for service in America.[6] They were to be commanded by Lieutenant-Generals von Heister and von Knyphausen, and there were a number of other experienced and efficient senior officers in their ranks, such as Major-General Baron von Riedesel, who was to play a prominent part with Burgoyne both in Canada and at Saratoga. According to an estimate laid before Parliament, with these Hessians (as they were usually termed), Canadians, Indians and Loyalists, there would be a total of some 59,000 men eventually available to fight for the Crown in America.[7]

Once the problem of procuring reinforcements had been solved, there was still that of transporting and supplying them which, of course, had to be by sea and up river. Transports could be blown off course, or captured by American privateers, and when the French entered the war, local command of the sea could not be relied upon, although fortunately, for the first two years, the French navy concentrated more on the West Indies than on the North American seaboard. The long journey was a limiting factor in every calculation, for by the time important dispatches arrived they had quite often been invalidated by events, similarly the uncertainty connected with the arrival of reinforcements and supplies could have an adverse bearing on strategy. In the case of supplies it was estimated that every soldier had to be supplied by sea with one third of a ton of food each year,[8] and every item of equipment.

There was a Commissary General (a civilian) with six deputies and assistants in charge of forage, cattle, fuel and brewery, and of course certain items of food were available in America – but spasmodically, and after much organising. These included grain, flour, rice, oats, hay, and fresh meat, but seldom in sufficient quantity at the right time in the right place. There never seemed to be enough ships, and more seriously, there was a lack of co-operation in London. Lord Sandwich, despite some obvious faults, did a very good job at the Admiralty, as did Lord Barrington at the War Office, but throughout the war each department of government fought its own corner independently, and only grudgingly surrendered ground to the war effort.

The British made some use of certain Indian tribes, but they were unreliable, difficult to manage and occasionally treacherous. On the other hand the Tories of America provided a considerable contribution to the British effort, and, as we have seen, were subjected to very rough treatment from the Patriots for their sustained loyalty and sacrifice. (The word Tory is synonymous with Loyalist, just as the Patriots sometimes called themselves Whigs.) In the early months of 1775, when war threatened, some provincial governors (especially the Southern ones) submitted wildly optimistic reports as to the number of eager men who could be counted upon to play a powerful part in suppressing rebellion. Such reports were misleading. Nevertheless, from the very beginning, and despite niggardly inducements to volunteer, there were men who by forming independent companies, or by joining the regular Army, helped fill the gap when numbers were badly needed.

When attempting to assess the numbers and types of persons who eventually became active supporters of the British it must be remembered that at the beginning of the war there were very few Americans with thoughts of independence. Most did battle for better conditions within the British Empire. Therefore not until the Declaration of Independence in July 1776 did many of those hitherto supporting the Patriots decide either to join the British, or at least to become neutral. The bulk of the active Loyalists came not from the diehard, thoroughly English Americans, but from those people who, while loving America, wished to be ruled from England, and who considered the extreme demands of the Patriots unnecessary. Probably one third of the population supported the British, and another third may have been neutral. The American historian Claude Van Tyne states that eventually there were more than 50,000 Loyalists throughout America, regular or militia in British service.[9] But this estimate is probably too high, and 30,000 would be nearer the mark. The Colonial Office records state that the number of Loyalist troops under arms *at any one time* was 3,257 in 1777, 7,348 in 1778, and much the same number in 1779.[10] Undoubtedly the enrolment of Loyalists depended to some extent on British successes and vigorous action from which, they did not get great encouragement.

The majority of these Loyalists preferred to fight in their own units, mostly as provincial militia, although a fair proportion joined the regular army, especially when improved pay and allowances were offered after General Howe had assumed overall command in 1775 (see Chapter 3 and Chronology). The Royal Fencible Americans and the Loyal Irish Volunteers came into existence during the siege of Boston, a whole regiment of Scottish Volunteers raised in New York in 1775 was sent to Canada, and a year later two particularly wild Loyalist companies known as The Loyal Greens, and The Tory Rangers were creating unnecessary mayhem on the Canadian border. In the South, particularly in irregular warfare, the Loyalists played a prominent part. The exploits of Colonel Tarleton's British Legion in the Carolinas will be recounted later, and it was here that the Macdonalds were active. Allan Macdonald of Kingsborough married Flora of 'Over the Sea to Skye' fame, and in 1774 they emigrated to North Carolina, where, on 10 January 1776, he was commissioned to raise an array, while Flora became an active recruiter for the Loyalists. Her enrolments came mostly from the large colony of Scottish emigrants in that province, who formed The Royal Emigrant Regiment.[11] Colonel John Simcoe's Queen's Rangers was another Loyalist force that campaigned vigorously in the South.

It is probable that Loyalists continued training with the provincial militia until almost the outbreak of war. A fair number of both Loyalists and Patriots had had considerable fighting experience in the French and Indian wars, but that had been ten years or more earlier, and it was not until the threat of rebellion became a realistic possibility that the militia was spurred to an activity completely lacking in the immediate past. Even in Massachusetts, the province closest to the centre of affairs, there had been little sense of urgency where the militia was concerned, and drill and exercises were carried out spasmodically and haphazardly without much enthusiasm.

By the middle of 1774, however, all this had been changed. New companies were now formed, drill and training put on a regular basis and great efforts were made to collect arms and equipment. All instruction had to be within the unit, for there were no military academies or staff colleges. Muskets and powder – particularly the latter – were always to be a serious problem for the Patriot army. Only a very little powder was made in the colonies, and the few gunsmiths could not hope to keep pace with demand. Most Americans possessed a rifle of sorts, which helped the arms situation, but this was offset by an irrepressible desire to waste powder and shot on every animal, and even bird, that crossed their path. Clothing too was in very short supply, for America was not a manufacturing country, and what passed for a uniform was usually a round hat, and a hunting shirt worn over trousers and belted at the waist. To begin with, officers were indistinguishable from their men. However, in 1776 they were ordered to wear coloured cockades in their hats. At least in the early days, the soldiers were well fed, although later there were periods of great hardship through shortages of meat and vegetables.

In theory, every man of military age was liable for service in the provincial militia and every town had a militia company. Terms of service for three months were ridiculously short, especially as the intended policy of the American army was to delay and wear out its opponents rather than to invite a general action, and generals were often embarrassed by the sudden disappearance of their militia men. At the other end of the scale from this foolish resolution was a Massachusetts innovation which laid down that at least a quarter of enlisted men should form a special force permanently on stand-by to march at a minute's notice with arms, and a fortnight's rations. This was adopted throughout the army, and the so-called minute-men were of great value.

Congress, sitting in Philadelphia, may have given birth to the Continental Army, but it was in Massachusetts that it was conceived. Greater attention was paid there than in any other province to the training and exercising of the local militia; the first engagements of the war were fought in that colony, and it was the efforts of their local congress that persuaded the neighbouring New England colonies to send contingents that made up the force of some 14,000 men who not only besieged Boston, but formed the nucleus of the Continental Army. (This was formed officially on virtually the same day as Washington was elected to command it, i.e. 15 June 1775. It was therefore formed two days before Bunker Hill, but after Lexington. It was formed, controlled, and paid by the First Continental Congress.) Those men were joined in July by contingents from Pennsylvania, Virginia and Maryland, bringing the besieging force up to between 16,000 and 19,000.[12] Most sources give this force as the largest number of men the American army ever had under arms at any one time, although Carrington writes of a figure in the region of 38,000.[13] Numbers cannot easily be verified, for although soldiers of the Continental Army signed on for three years or the duration of the war, the army existed on enthusiasm and persuasion, certainly not on discipline, and men were inclined to come and go.

Thus, very briefly, the existing set-up and future prospects for the opposing armies. Meanwhile, in the early months of 1775, General Gage in Boston had become increasingly depressed at the deteriorating situation in America generally, and in Massachusetts in particular. He spared no effort to communicate his forebodings to the government, who persisted in blinking the facts, and failing to face realities. There were very few in England who wanted war, but many who still thought the colonists could be brought to heel by firm and repressive measures. In consequence, Gage's constant appeals for many more soldiers went largely unfulfilled, although limited reinforcements were promised for May. Meanwhile a dispatch from Lord Dartmouth (North's step-brother and Secretary of State for the Colonies), which reached Gage in April, urged him to take strong measures at once before the rebels could become organised.[14]

There was sense in this, but unfortunately, like so many ministerial decisions, it was made too late, for the Patriots were very much more organised than the government realised. The Massachusetts Provincial Congress, sitting in

Cambridge, had been active in the enrolment of militia men, and now had some 12,000 drilling openly, nor had its appeal to neighbouring colonies gone unanswered. The Committee of Safety (one of many committees existing at this time) had also been at work in procuring a fair quantity of powder and cannon. However, in this instance numbers were not everything, and Gage reckoned he had a considerable advantage in the training and discipline of his small force, which consisted of five weak brigades, commanded, in numerical order, by Brigadier-Generals Lord Percy, Robertson, Pigot, Jones and Grant.[15]

Gage was also instructed to arrest the leaders of the Provincial Congress (presumably on a charge of inciting rebellion), but that was a particularly foolish measure, for it would only create martyrs, and replacements would soon be found. However, he felt he should make a gesture by trying to catch up with Samuel Adams and John Hancock, who happened to be in the Concord area where a quantity of military hardware was stored. These and other caches of powder located at Charlestown and Salem had always been his target, and he now hoped that, with strict secrecy, he might snatch Adams and Hancock from the jaws of the militia at the same time as the raid on Concord. However, Gage had yet to learn that when operating among a predominantly hostile population secrets were difficult to keep.

Concord is some 20 miles to the north-west of Boston, and to accomplish his purpose Gage sent the flank companies* of the Boston garrison across the harbour by boat, then up the Charles River to what is now Lechmere Point, and from there by a night march to their destination. Only Gage, Colonel Smith of the 10th Regiment, who was to command the 750-strong force, and his second-in-command, a thrusting major of Marines called John Pitcairn, had knowledge of the plan, and no one was allowed to leave Boston. But spies were alert, and troop movements are useful indicators; soon the legendary Paul Revere and William Dawes succeeded in leaving the town (the former by boat to Charlestown, the latter through Roxbury), and riding through the night got word to Adams and Hancock, asleep in a village near Lexington, that the redcoats were on their way. In the early hours of 19 April the two congressmen rode to safety, and shortly afterwards Pitcairn, commanding an advanced detachment, reached the village. Their march during the night had gone unhindered, but not unseen, for Revere had roused the militia as he rode, and the alarm bells rang and guns were fired.

At Lexington Pitcairn and his men found their way barred by some 70 militia under John Parker. There was a brief confrontation and some shooting – no one knows who fired first – and eight Americans were killed before Parker broke off the engagement. Pitcairn's party was then joined by the main body, and they marched to Concord where they arrived at about 7am. Here they destroyed what few military stores the Patriots had not already removed,

---

*The grenadier and light infantry companies of the Boston regiments had been massed to form seperate grenadier and light infantry battalions.

before becoming embroiled with at least 400 militia. There were some casualties on both sides before Smith decided it was time to withdraw.

It was 20 miles back to Boston, and almost every one of them was to be stained with British blood. As far as Lexington the retreat of these very weary soldiers was fairly orderly, but the countryside was alive with Patriots crowding in on their exhausted quarry hoping for a kill, and retreat became a disorderly rout through the town. The situation was only saved by the arrival of Lord Percy's brigade of some 1,200 men. Two thirds of the entire Boston garrison were now exposed to the accurate sniping and hit-and-run tactics not only of the militia but of an ever increasing mass of local men armed with a wide variety of weapons. Percy's brigade, itself by no means fresh after a rapid march, shepherded Smith's men as best it could with strong flanking protection, and the use of its cannon. But the march was still something of a shambles with the British at bay, yet finding time to destroy much property along their path. By the time the causeway was reached, and the troops were able to rest on Bunker Hill under the protection of the ships' guns, they had lost 73 killed and some 200 wounded. The American casualties were only 95, of whom 49 were killed.[16]

The British had seriously underestimated the courage, resolution and ability of the Patriots. In this dress rehearsal for bigger operations not much had been learnt: there were some hard lessons still to come, and in Massachusetts to come very soon. But for the Americans Lexington was a propaganda gift, and every advantage was taken of it in a wildly extravagant manner. The news spread through the colonies at almost the speed of light with grossly exaggerated accounts of British atrocities. Morale rose steeply among the Patriots, for it could be seen that, given the type of country to be fought over, the British were certainly not invincible. The general euphoria was shared by members of the Second Continental Congress, which on 10 May debated Lord North's conciliatory measures, and it was no surprise in the circumstances that members declared these did not go far enough to meet American demands. War had come a stage closer.

As a direct result of Lexington and Concord, provincial congresses sprang into action, and quickly organised local militia. Foremost of these was Massachusetts, which stirred by a forceful appeal from Dr Joseph Warren, the greatly respected President of the Massachusetts Provincial Congress, soon had many eager recruits to swell the number already undergoing training. They were mostly farmers, and fairly rustic, nor was it easy to supply them with powder, but they were in sufficient numbers to keep Gage and his men penned up in the Boston peninsula.

From his vantage point on Beacon Hill, that general looked anxiously on the camp fires of an encircling army from north of Cambridge southwards to Roxbury. He had fortified Boston Neck in 1774, but he had made no move to occupy the two vital features that overlooked Boston, Bunker Hill to the north on the Charlestown peninsula, and Dorchester Heights across the water to the south-east. Having missed the opportunity early on he would find it increasingly difficult to make the attempt, nor for that matter would it be easy for him to force a break-out. However, at the end of May seven battalions of infantry and the 17th Light Dragoons arrived giving Gage an effective fighting force of almost 8,000 men, although he was very short of horses and wagons. Embarkation of these reinforcements was so delayed that when an officer was sent to New York to buy horses for the dragoons the town was found to be in the hands of the revolutionary party, and horses could not be obtained.

The new troops sailed into Boston harbour on 25 May. A few days earlier the *Cerberus* had arrived bringing three major generals. They were William Howe, Henry Clinton and John Burgoyne. All were destined to play leading roles in the forthcoming war although the reason for their immediate presence is not entirely clear. Howe, who was the younger brother of Lord Howe, soon to command the fleet in American waters, had considerable military experience, including fighting at Quebec under Wolfe, who had formed a very high opinion of him. He had been a major general for three years, and was respected as a strict disciplinarian with a cool, incisive mind which, when applied, produced good results. But he was indolent and delighted to dally. His overall performance in America was to be disappointing. Clinton, at 45, was the same age as Howe, whom he was to succeed three years later. He had seen action in the Seven Years' War, and had spent some time in America when his father was Governor of New York. He was never entirely certain of himself, or of others; he was also a compulsive complainer, and inclined to shy away from responsibility. Nevertheless, he had considerable tactical ability, although insufficient to justify his appointment as commander-in-chief in a difficult theatre of war.

The third general, John Burgoyne was, at 52, seven years older than the other two but junior in service. He too had had considerable military experience. Originally in the Horse Guards, he fought in Flanders with the 1st Royal Dragoons and had also seen action in Portugal at the head of his own regiment, the 16th Light Dragoons, which he had been commissioned to raise in 1759. Like Howe, Burgoyne was a Member of Parliament, and like him had expressed hopes that the American problem could be resolved by consultation rather than war, and neither of them was over keen to serve in America. Burgoyne was also a dramatist with, later on, two or three good plays to his name. His languid courtier air was not unknown at the gaming tables of London's clubland. It did not take these generals long to appreciate the need for offensive action before the ring could be tightened even more.

Artemas Ward, an elderly major general of the Massachusetts militia, had been given command of the rapidly assembling troops, and he had established his headquarters at Cambridge, which was the central rallying point for the incoming volunteers. The original British plan was to launch an attack on Cambridge on 18 June, with subsidiary ventures to seize and occupy Bunker Hill and Dorchester Heights.[17] However, little could be kept secret in Boston, and the Patriots soon got wind of the plan. The Committee of Safety, in

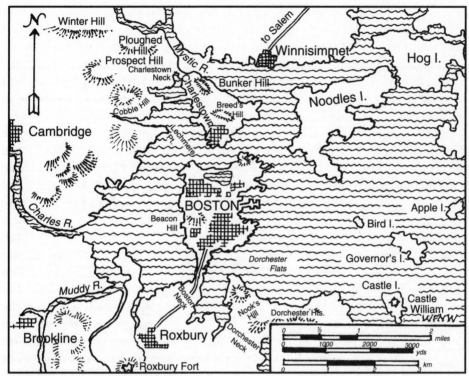

Boston and its Environs

conjunction with the Council of War, therefore voted to send a force of some 1,200 men to steal a march on Gage and occupy Bunker Hill.

The Charlestown peninsula is a roughly shaped triangle with a causeway at its apex, which connects it with the mainland and forms a barrier between the Mystic River to the east and the Charles River to the west. At the north-eastern end of the peninsula is Bunker Hill, a feature some 110 feet above sea level with the ground sloping gently from its base for a little under half a mile, only to rise again with another lesser feature of 75 feet known as Breed's Hill. Just south of this hill was Charlestown, a small place of about 3–400 houses, most of whose inhabitants had departed under threat of British naval guns.

The principal men on the American side concerned in the battle of Bunker Hill were Generals Ward and Putnam, Dr Warren and Colonels Prescott and Gridley. Ward and Warren were not in favour of fighting the British on such an exposed position with a shortage of ammunition, but the others were for battle and when Warren gave way (and forsook his presidential position for the humbler role of rifleman) it was decided to proceed, and the command was given to Colonel Prescott, a farmer by trade but with active service experience in the French wars. Colonel Gridley, who was his Chief Engineer, had been in the British army under Wolfe, but for some time he had supported the Patriots.

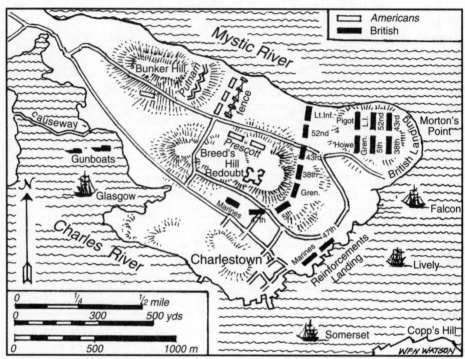

Bunker Hill and Breed's Hill

On the evening of 16 June Prescott led his men (mostly plain clothed) out of Cambridge with a good supply of entrenching tools, but with horns and pouches dangerously short of powder and shot. He marched over Bunker Hill to take up a position on the lesser feature, Breed's Hill. For this he has been criticised, because it was just within range of British naval guns (although in the event they did little damage), and because being so far forward the force could be cut off. Had he received promptly the reinforcements he called for, and been able to prepare a fall-back position on Bunker Hill, his choice would undoubtedly have been correct. As it was his men toiled all through the night, and when morning came although the position was not completed it was far enough advanced to serve its purpose.

On the hill's summit a fairly serviceable redoubt had been erected with a solid face 45 yards long. Stretching back from the redoubt, and at a sharp angle, was a short breastwork, and from that an existing hedge and stone fence ran some 200 yards behind the redoubt almost to the little Mystic River near the sea. This natural obstacle had been strengthened and made into a double line. The whole position gave adequate frontal protection, and to guard his right flank and to harass any frontal attack Prescott placed some men in Charlestown. However, until Ward sanctioned the summoning of Colonel Stark's famous New Hampshire regiment, there was a dangerous gap of 200 yards on the left that the British would seek to turn.

Gridley had done a good job in laying out the position, and the rough untutored patriot soldiers had done an even better one in preparing it to resist attack in the few hours given them. But it had taken the edge off their fitness for battle, and they were short of food and water, neither of which was forthcoming on that dawn of the 17th or later. It was yet another breakdown of the headquarters administration, and affected morale to the extent that some men discarded their entrenching tools and slunk away, but the majority (between 1,500 and 1,700) stood firm to face the first assault which Howe, who was in charge of the British force, was about to launch.

It had come as a considerable surprise to Gage to see the Patriots established on Breed's Hill. Some of his sentries had heard the digging during the night of the 16th, but apparently had not reported it. Any attack on Cambridge, or the seizure of Dorchester Heights had now to be abandoned. However, Gage ordered his gunners to put down a heavy concentration of shot and shell on the Roxbury defences to foil any thoughts General Thomas, commanding the Patriots in that area, might have of attempting an attack on Boston neck or of sending aid to Prescott.

The British troops came across the narrow strip of water that divided the Boston and Charlestown peninsulas in barges to land at Moulton's Point in the south-east of the peninsula. The June weather was extremely hot, and they undoubtedly suffered greatly with rolled blankets atop thick scarlet tunics, and packs (which they did not discard for the first two attacks) stuffed with three days' rations. As each wave landed, the barges returned for more and the troops deployed on Morton's Hill. Burgoyne says there were about 2,000 men in the initial landing,[18] which was carried out with perfect precision and drill discipline, and supported by the batteries firing from Copp's Hill and the ships. Gage had the support of three ships of the line (*Somerset*, 68 guns; *Glasgow*, 34 guns; and *Lively*, 20 guns) and two gunboats. The bombardment on the redoubt had little effect, but the covering fire ensured that the landing and forming-up were unopposed.

The attack was to be carried out by 10 companies of grenadiers, 10 of light infantry and two battalions, but no sooner had Howe and his second-in-command General Pigot landed, than they realised that their force was insufficient for the task ahead, and Howe sent at once for two more battalions. Clinton had strongly advised landing a force behind the enemy defences to strengthen the attack, and cut off the retreat;[19] but Gage would have none of it, being quite certain that a frontal attack would quickly dislodge the Patriot rabble from behind their hastily improvised defences. And so Howe eventually formed up for the attack in column of battalions with 10 companies of grenadiers, 10 of light infantry, the 5th, 38th, 43rd and 52nd Foot. Howe himself commanded the main attack on the redoubt while Pigot attacked to his right on the American left. There are numerous accounts of Bunker Hill, and almost all of them differ in regard to the British order of battle. But it seems certain that the regiments mentioned above were in the first attack, and that the reinforcements which arrived with Clinton were a battalion of Marines and the 47th Foot. What is not certain is the composition of the right and left columns in the first attack. Howe had the grenadier and Pigot the light infantry companies, and it seems quite likely that Pigot had the 43rd and 52nd, leaving Howe with the 5th and 38th.

Howe began his attack at about 3pm on this very hot afternoon, and as the troops trudged under their heavy packs through the long grass and broken ground of Breed's Hill the Patriots silently watched and waited. But soon the stillness of the afternoon was shattered by the rumbling thunder of iron shot as Burgoyne directed a lively cannonade from Copp's Hill, which was supported by the gunboats' floating batteries. Burgoyne's letter says that as Howe's 'first Arm was advancing up the Hill, they met with a thousand impediments from strong fences and were much exposed. They were also extremely hurt by muskets from *the town* of Charles Town.' Howe 'sent word by boat' to Burgoyne asking him to set fire to the town. 'No sooner said than done.' writes Burgoyne, 'We threw a parcel of shells & the whole was instantly in flames.'[20]

In this first attack it would seem that Howe's principal aim was to turn the American left with Pigot's men and roll it up, but by the time the attack began, the gap in the line between the end of the defences and the river was no longer there, for Stark's New Hampshire men had arived to join others from Connecticut. All along the line the Americans held their fire; the British columns, on the other hand, had no sooner deployed at about 100 yards from their objective than they opened a random fire at too great a range. With considerable skill Prescott and his officers restrained the Patriots until the British were almost on them, and then they delivered a relentless volley of musket fire, all the more damaging as many of their old weapons fired a load containing one bullet and anything from three to nine buckshot. The redcoats were mown down in tangled heaps by this scything fire, and were ill served by their field artillery, because through some grave error the wrong calibre shot had been sent forward. The men staggered and faltered, and although encouraged by the heroic example of their officers and sergeants, they could not get close enough to use the bayonet, and were forced to fall back.

The battlefield formed a natural amphitheatre, and from the high ground in Boston a vast gathering was able to watch the British battalions reform, and go into action again. As before, they deployed along the whole line, and as before they were met with a withering fire that further decimated their ranks. But by now there were fewer officers, for gold lace and epaulettes made splendid targets for specially selected sharpshooters and at one time Howe, who gallantly led both assaults, found himself virtually isolated in front, all his staff having become casualties. Again the British troops reeled and retired back to the shoreline.

On the American front morale was now high, for these rustic soldiers had gained confidence in their weapons, and their own skill in handling them. General Putnam had ridden the battlefield animating his men, and trying to hustle reinforcements over the causeway and up to the front. But in this he was unsuccessful, for *Glasgow* and the gunboats had the range, and so it was mainly the men who had built the defences who had to bear the brunt of the fighting. Far worse than the failure to reinforce was the lack of ammunition. Officers had done all they could, and with considerable success, to impress a strict fire discipline; even so there were now many men with no more than three rounds in their pouches.

Meanwhile, Howe wasted little time in rallying his soldiers for yet another scramble up the hill, because he knew well that after two bloody repulses any lengthy delay would make it difficult to galvanise even veterans for a third dose of this lethal mixture. Fortunately Clinton, who had been one of those watching the earlier strife and slaughter, realised the need for more men. In very quick time, and without any orders from Gage, he managed to collect some 500 reinforcements to follow him over the water.[21] Helped by these fresh troops Howe again led the way forward; this time there had been sufficient sense for orders to be given to remove packs, and the main thrust of this third attack was to be against the redoubt and principal defences. The light infantry companies on the right were merely to hold the American left.

For a short time it looked like being a third disastrous rebuff, for the Patriots, holding their fire until their enemy was within 20 yards, made every shot count, but there was nothing left in the pouches and when the British were at last able to storm the defences the bayonet settled the issue, for the Americans had none of these deadly weapons. At almost the very end two men, one on either side, were killed who could ill be spared. Major Pitcairn, a most promising officer who had distinguished himself at Concord, was shot as he entered the redoubt and Dr Warren, who had given so much to the Patriot cause and who, with Colonel Prescott, was among the last to leave the redoubt, received a bullet in the head. With the capture of the redoubt, the battle was not completely over, for Stark's men at the Mystic River, and a party of newly arrived troops at Bunker Hill under a Major Jackson, fought a stout rearguard action until the last Patriot had crossed the causeway. Putnam himself had worked hard to rally sufficient men to establish a position on Bunker Hill, but although there was no organised pursuit this had proved impossible, and the day ended with him occupying the field works on Prospect Hill.[22]

Howe claimed the victory, for he occupied the battlefield. However, as the casualties showed it was a very hard won victory, and certainly not one to celebrate. As always, figures vary considerably in different accounts; Gage's detailed official account, sent to Dartmouth on 25 June, gives the total of all ranks killed and wounded as 1,741, which included 24 officers killed. That is

more than 50 per cent of the force actually engaged, and is considered too high
by Fortescue and most other historians who think a figure of a little over 1,000
to be more realistic, but even that is almost 40 per cent of the force engaged.[23]
Burgoyne correctly wrote, 'The loss was uncommon in officers for the number
engaged.'[24] The American casualties were given as only 450 killed and
wounded, all ranks.[25] They had good reason to be proud of the way their men,
many of them raw and quite untrained, had stood the test of battle.

* * *

The whole affair of Lexington and Bunker Hill was a most inauspicious
beginning to the military operations that were to take place on the American
continent over the next seven years. It is some consolation that most of the
mistakes and miscalculations that occurred were committed by General Gage
who was not to have the chance to make any further errors of the kind. But at
least he had been proved right in his request for 20,000 men, for if Bunker Hill
had shown nothing else, it had made it very clear that the Patriot forces,
although still untrained and undisciplined, could fight with courage, determination and a modicum of skill. It might have been thought that an initial
hopeless underestimation of the enemy would have been cured by this battle,
but it was to take many politicians in London and senior officers in America a
very long time to appreciate the remarkably high qualities of their foe.

Another dangerous mistake that was to persist for over a year was a fatal
indecision on policy between suppression and conciliation, which inevitably led
to half-measures such as the bungled attempt to arrest Adams and Hancock,
and ambiguous orders to commanders in the field. Again, considering how
much value was attached by the government to the use of Loyalists, recruiting
by the army had been only half-hearted. With greater effort and the offer of
better conditions, more volunteers would almost certainly have come forward
at this most important early stage. As reinforcements from home were always
to be uncertain, and somewhat haphazard in timing, early enrolment of
Loyalists was vital, not only for numbers but even more importantly for
example.

The actual military mistakes began well before the battle, and were of course
made by Gage before the arrival of the generals from England. Presumably
there had been no proper reconnaissance of the surrounding area, or else
surely Gage would have realised the great importance of the Dorchester
Heights and Breed's Hill, and would have occupied these key positions. Failure
to do so was the cause of considerable trouble to himself and others. Moreover,
his intelligence service would seem to have been almost non-existant, for the
movement of troops on to the Charlestown peninsula, and the entrenchments
on Breed's Hill, came as a complete surprise and put him at a great
disadvantage.

Technically Howe was correct in claiming the battle as a British victory, but there had been much disorder and confusion in the ranks which was probably the reason why the victory was not much more decisive. The heavy casualties Howe had to accept were serious enough in themselves, but they seem to have had an adverse effect on his future performance, for it was noticeable that on occasions he was to err unnecessarily on the side of caution. No one can impugn his courage or his leadership, although his reason for letting the soldiers make two attacks heavily burdened with packs will always be a mystery.

But of the generals it was Clinton who showed strategical sense and initiative, for his advice to land a force, covered by the guns of the fleet, behind the enemy's defences and to occupy Bunker Hill would have completley cut off Prescott. However Gage, presumably with Howe's concurrence, insisted on only a frontal attack with its dire consequences. And it was Clinton's initiative that virtually won the day, for without the extra 500 fresh men, the original force, which had fought with such enormous courage and perseverance, might never have gained the hill.

Burgoyne played no active part other than to destroy Charlestown. He seems to have been well pleased with the battle, for in his letter home he wrote, 'The day ended with glory, & the success was most important, considering the Ascendency it gave the regular troops.'[26] With great respect to my ancestor, that is nonsense, for if there was an ascendency as a result of the battle, it belonged to the Americans.

At the end of the day there was no pursuit. Although the American retreat had been more orderly than might have been expected from raw troops not long under discipline, the men were tired, dispirited and virtually without any ammunition. Howe had troops available, who not having had to struggle up Breed's Hill twice with heavy packs, were comparatively fresh, but he held them in leash, for he never was a champion of the pursuit. At Bunker Hill there had been every chance to destroy the Patriot force, and if this had been accomplished, and the besiegers driven from Boston (perfectly possible) further bloodshed might well have been avoided. But it was not to be, and through a series of blunders great and small a great opportunity had been lost.

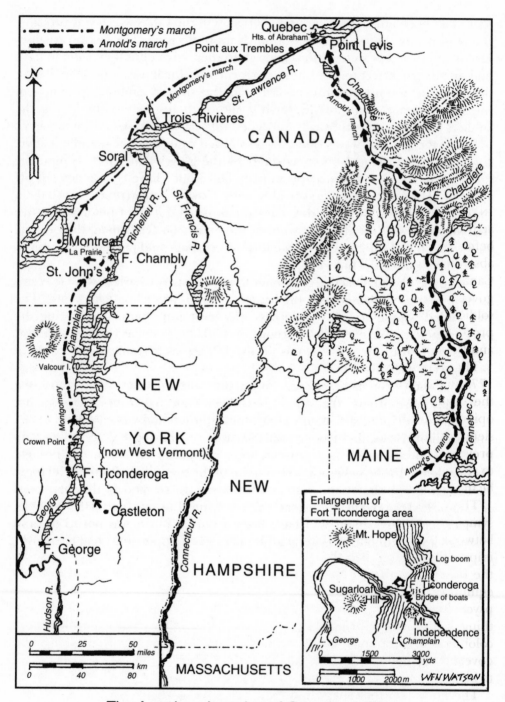

The American Invasion of Canada, 1775-6

CHAPTER 3

Initiative and Inaction

IN THE REMAINING months of 1775 after Bunker Hill the principal military activity concerned the American attack on Canada, which spilled into 1776 before being finally halted just south of the border. It was ill-conceived and had absolutely no future, but it came very close to succeeding. The British troops in Canada were under the command of General Sir Guy Carleton who had been governor and commander-in-chief since 1766. He was an extremely competent soldier with considerable fighting experience, and he was an excellent proconsul who had gained the respect of most of his peoples – English, French and Indians.

The American plan to invade had been initiated in Congress before 15 June 1775, the date upon which George Washington had been elected to command the Continental Army. At first he disapproved of it, but later altered his opinion when he thought he saw possibilities which seemed to many to be very slender. Strategically there was sound sense in occupying the frontier forts to prevent an invasion down the Hudson to cut off the New England colonies, but to risk an invasion of Canada with the Patriot army as it then was would seem to be the height of folly.

There was no evidence to support the likelihood of the Canadians rising in sympathy with the New Englanders, and there was no particular empathy between the two peoples; the direct approach to Quebec was through almost impossible country; had that city been captured it could not have been held in the face of the British Navy; the Americans had neither the troops nor the equipment to spare for such an enterprise; and finally it could have been a psychological error. To rebel against oppression in their own country was one thing, but to invade British territory was quite another. At a time when most colonists still entertained hopes (false as it happened) that the British king and government might relax their stubborn attitude, such action was not calculated to help.

The whole idea took shape in the autumn of 1774 when the First Continental Congress made unsuccessful overtures to the Canadians for support and co-operation. This was followed by the daring reconnaissance of a lawyer called John Brown who returned with encouraging information as to the attitude of many Canadians and Indians in the event of an invasion, and the fact that

Carleton, who had to send troops to Boston, had only 'about 400 regulars, besides the garrison of 300 at St Johns'[1] with which to defend the province. As is so often the case, composition and numbers differ somewhat in each account (See source note). Brown's favourite report triggered American action, from which the initial impulse came from Benedict Arnold, the man to prove the best fighting general on both sides in the war. Sadly, his personal daring, tactical boldness and skill in battle are tarnished for us by an unreliable and unprincipled character, and the odium of military treachery (as we shall see in Chapter 10). Now, in May 1775, while commanding a company of the Governor's Guard in New Haven, he absented himself without leave and turned up at Cambridge with a proposal to take Ticonderoga, a very important strategically sited fort situated between Lakes Champlain and George. The Committee of Safety promptly conferred the rank of Colonel on him, and authorised him to enrol 400 men for the task.

Simultaneously, another equally impetuous freelance soldier called Ethan Allen had the same idea, and putting himself at the head of a particularly tough band of farmers known as the Green Mountain Boys he rallied at Castleton in what is now Western Vermont. Here Arnold caught up with him and tried to pull rank, but when the Boys clearly showed they would not serve under Arnold a joint command was agreed upon.

Ticonderoga, built by the French in 1755, and called by them Fort Carillon, is strategically placed between Lakes Champlain and George. It had fallen into disrepair, and Gage had not sufficient troops to garrison it properly, but he had taken the precaution to warn its commander, Captain William De la Place, to be on his guard[2] and had sent an urgent letter to Carleton asking him to despatch a regiment either to that fort or to neighbouring Crown Point. The request reached Carleton too late for action and De la Place, although in command of only some 40 soldiers, apparently failed even to guard against surprise. The Americans rushed the building in the dead of night and in a matter of minutes, without a shot being fired, had rounded up the garrison and secured the fort. The next day (11 May) the few men in Crown Point surrendered at once to Allen. A number of cannon, ammunition and stores were soon available for despatch to Cambridge where they were badly needed, and Carleton's route south from Canada was blocked. A small, but important affair, and not one to the credit of British arms.

The tiny British garrison at Fort St John, just across the Canadian border, soon fell prey to Allen's Green Mountain Boys, but could not be held in the face of rapidly approaching British reinforcements from Montreal. Meanwhile Arnold, who had not partaken in this latest coup, had had his second attempt to assume overall command rejected, and so dismissing his men he had returned to Cambridge somewhat out of temper. But not before he had written to the Continental Congress to say that in view of Carleton's shortage of troops, and his own knowledge of Quebec and its environments, he could march 2,000 men up the Kennebec and Chaudiere rivers, traversing well over 100 miles of complete wilderness, and take Quebec. It was a typical piece of Arnold boastfulness; nevertheless when Washington later took him at his word, he had sufficient guts and resolution almost to fulfil his boast.[3]

Any plan to take Canada required the capture and occupation of Montreal as well as Quebec and, animated by Arnold's confidence, a two-pronged attack was to be made. Three thousand men would rendezvous in August at Ticonderoga and have Montreal as their first objective, while Arnold was given 1,100 men to march across the wilderness to Quebec. General Philip Schuyler was to command the Montreal force, with General Richard Montgomery as his second-in-command. Allen and Seth Warner (who had succeeded Allen in command of the Green Mountain Boys) marched with Schuyler. The latter was not fit and, perhaps because of this, he dithered and procrastinated giving Carleton precious time to organise his defence. Had Montgomery been put in command initially (he took over when Schuyler went sick in September), Carleton might have had difficulty in saving Canada. Montgomery, an Irishman who had served in the British Army, was a very talented and energetic commander.

Montgomery was held up for 50 days by the garrison of St John's, now increased to 500 men under Major Charles Preston, for he was not strong enough either to assault the fort or to besiege it and march on. He was short of almost everything, and indeed might never have succeeded had not Fort Chambly, about 10 miles to the north, been disgracefully surrendered after only a day and a half of siege. This gave Montgomery all the cannon, powder and food that he needed to force Preston to capitulate.[4] The latter, and his 500 men, had quite possibly saved Canada. The defenders of Montreal, who had defeated an unauthorised and impulsive attack by Allen, whom they captured, opened the gates of their undefended town to the conqueror of St John's on 11 November. Carleton had worked hard to collect a motley force of 800 Canadians, Indians and regulars with which to supplement the small numbers in Montreal, but Warner and the Boys had scattered them on the banks of the St Lawrence, and Carleton only narrowly escaped capture himself on his way back to Quebec.

Meanwhile, Arnold, with his 1,100 men, and some 200 rough-hewn and leaky batteaux, started up the Kennebec on 25 September.[5] The conditions they had to contend with were appalling, for almost all of their journey was through primeval forest. When they were not navigating their boats through green mats of sinuous waterweed or manhandling them up hills, they had to hack their way through thick undergrowth as dense as tropical jungle and littered with fallen forest leviathans, and scramble along narrow moose tracks overhung with more dense green undergrowth. The weather was ghastly, and the rations meagre. Only the toughest survived; three companies of militia turned back, and almost 100 men fed the wolves. No more than 675 weary and bedraggled soldiers reached the St Lawrence at Point Levis on 10 November,[6] Here Arnold had to seek boats, but such was the leadership of this man that by

14 November he had what remained of his force across the river and on the Heights of Abraham covering Quebec and ready to assault.

Before Carleton got back to Quebec, Colonel Alan McLean, himself not long back from Montreal, set about strenthening the defences with skill and energy. Carleton had scraped up about 1,800 men with whom to defend the city, and he was better supplied with arms and ammunition than were the Americans. Arnold, impatient as ever and with typical sauciness, summoned the city to surrender. After the inevitable refusal, he withdrew his troops to Point aux Trembles to await Montgomery, who arrived on 2 December. The two forces, numbering some 1,200 men, then invested the city and awaited an opportunity to assault. This came on the last night of 1775. The attack was made in a blinding snowstorm from two separate points, but surprise was lacking and in neither case was success achieved against a resolute defence. Montgomery was killed, and Arnold severely wounded in the leg. Any further assault was impossible, nevertheless the wounded Arnold hung on tenaciously throughout the winter with a close investment.

Thus it remained for almost six months, although on account of his wound Arnold was superseded in command first by General Wooster then by General Thomas, who arrived before Quebec on 1 May 1776. The troops that had fought their way into Canada had been reinforced by several companies of militia, but Thomas found an army sadly depleted by disease and starvation, there being scarcely 500 men fit for duty. Carleton could probably have destroyed them long since, but he preferred caution, well aware that, come the spring, he would be reinforced. And on 6 May, three ships forced their way through the ice bringing the necessary troops for Carleton to begin his slow but inexorable offensive to drive the Americans out of Canada.

At first the retreat was precipitous, for the Patriots were neither numerically nor physically in a position to stem the onslaught of fresh British troops. But then Washington, somewhat reluctantly, sent 3,000 reinforcements under General John Sullivan who, unaware that Carleton had received a further reinforcement (which included General Burgoyne), attacked the British line at Trois Rivières. In spite of the additional troops, Thomas, who led the attack, was well beaten and retired some 25 miles to Sorel. Here he paused until the British fleet was within an hour's sailing of that place, and then he began a steady withdrawal back to Crown Point. Arnold, still suffering from his wound but inexhaustible, headed 300 men who hung on to Montreal until the ships were within a few miles of the town, then withdrew across the river to La Prairie. When he crossed the border in late June 1776 the last of the Patriots had been driven from Canada.

However, the northern campaign was not yet over. If the British were to advance into America down the Hudson, it was necessary to have command of Lake Champlain, and both sides therefore began building suitable vessels for the forthcoming naval battle. The Americans were now under Generals Gates and Schuyler (recovered from his illness), but the moving spirit of their rush to build a fleet was Arnold. He was still in some pain, but most anxious to display his aptitude for organisation and nautical skills, which were based on a fair degree of sailing experience.

Carelton had been in no hurry after Trois Rivières, and did not respond to Burgoyne's urgency for a closer pursuit. There was said to be further friction between the two when Carleton vetoed Burgoyne's plan to take a mixed force of British and Indians (whose chiefs he had met in Montreal) up the St Lawrence to Oswego, and thence along the Mohawk Valley to get behind Gates and cut him off. Burgoyne, who was always a little inclined to think he knew best, and who was quick to come to conclusions and commit them to paper, had doubts as to Carleton's fitness for the command.[7]

Both armies got their mixed fleets of schooners, sloops and flat-bottomed barges ready for the water at about the same time, and soon one of those old-fashioned naval battles with ships coming alongside and troops discharging muskets, was fought on 11 and 13 October. But Arnold's squadron was outgunned, and off Valcour Island it was virtually destroyed, although he himself and most of his men survived the encounter. Carleton now had control of Lake Champlain, and held open the water communication down the Hudson. He felt that he had done enough, and that it was too late in the season to attempt an advance to Albany or even New York. For that decision he has been criticised, for there were those who said it could have ended the rebellion.[8]

Such criticism was probably unjustified, but he should have consolidated his position by taking Crown Point, which would have made a useful springboard for operations the next year. There is a letter, dated 26 October 1775, from one of his staff (Colonel G. Christie) disparaging Carleton's conduct of the campaign, and in particular criticising him for not pressing on to Ticonderoga, and saying it would have been abandoned by the enemy at his appearance. The letter ends with an opinion that Carleton 'is totally unfit for such a command', but not much weight should be attached to that for the writer, a seemingly unpleasant staff officer, bore Carleton a grudge for alleged 'ill-treatment' when he [Christie] was employed to contract a number of workmen for government service.[9]

It would be wrong to say the British blundered in this northern campaign, for that is too harsh a word, but there were some quite serious mistakes which fortunately for them had no lasting consequences. Ticonderoga was at the centre of what controversy there was. This stone, star-shaped fort with its five bastions stands at the end of a promontory jutting into Lake Champlain, and was strategically of considerable importance. It is easy with hindsight to say that as soon as trouble loomed Gage should have had it put into repair, well provisioned and well garrisoned with the dominating Mount Hope and Sugarloaf Hill (called Mount Defiance) features occupied, but he was desperately short of troops. Nevertheless, forty men was something and nothing, and Gage should have known that had he been able to hold

Ticonderoga, any attack on Canada would be aborted before it had begun. Similarly Carleton, nine months later, should have realized the advantage to be had in retaking this fort.

Carleton had conducted the defence of Quebec in difficult circumstances with great skill. He was badly let down at Fort Chambly where the surrender enabled the Americans to take St John's with its valuable booty. Fortunately, for Carleton the garrison there held on long enough to save Quebec. Burgoyne's adverse comments on Carleton need to be viewed with caution, for he was inclined to see himself as a better and more suitable general or governor than the incumbent he was criticising. It is true that had Carleton moved more quickly after Trois Rivières, he might have caught and destroyed Sullivan on the Richelieu river, but the Mohawk Valley project would have been a bold venture, which Burgoyne was to try later in different circumstances without success. Carleton was almost certainly right to reject it.

It is impossible to say whether the Americans would have taken Canada had they begun their campaign earlier and not split their force. Certainly they would have had a much better chance if the attack had got underway soon after the plan had been finalised, say in July 1775. They would then have had more time before Carleton received reinforcements.

It must have been a mistake to let Arnold waste men tramping through the Canadian forests; concentration of force would have been better. With more men, Montgomery could well have succeeded. It would, of course, have been only a temporary success, for the Americans could never have held the place, but reconquest might have been tiresome, time-consuming and costly for the British. However, it was an achievement that the Patriots had set back British north-south operations for more than a year, for the longer the delay the less chance these had of success.

* * *

At the end of June 1775, George Washington, now in command of the Continental Army, came to Cambridge. He had been chosen for that command on 15 June but his election had not been unanimous, for there were members of Congress who thought that the Army should be commanded by a Massachusetts man. But Washington was a natural choice. Aged 43, he was tall, slim and physically powerful; a man of great dignity, possessed of considerable confidence and tenacity of purpose. He had had much military experience in the French and Indian wars, notably when in 1755 he had served on the personal staff of General Braddock. During the fatal attack on Fort Duquesne (now Pittsburg), he took command of the General's force when the latter was mortally wounded. In the following year, he was given command of all the Virginian forces. Despite his many successes, Washington remained a modest man, entirely unencumbered with any sense of self-importance.

What Washington saw when he inspected the army besieging Boston displeased him enormously. Almost everything imaginable was wrong with his troops. To begin with, from an original 16,000 (19,000 according to Fiske) his army had dwindled to little more than 10,000 men, and these had very little powder and shot; much needed cannon was yet to be taken from Ticonderoga; discipline did not exist. The soldiers did what they liked, when they liked, were ill-clad and grossly insubordinate, moreover some of the officers were little better. Washington very quickly showed his mettle. With the full support of Congress, he took immediate steps to tighten discipline, where necessary through courts martial and the gallows. He got some sort of cleanliness into the camp, and issued a series of urgent inventories for stores of every kind. Rations at this time were plentiful, but were being unfairly and unequally distributed, which also he remedied.

General Putnam had occupied Prospect Hill before the fighting in Charlestown, and Washington now took steps to develop its advanced flanking posts of Lechmere Point, Cobble and Ploughed Hills, as well as ordering earthworks, redoubts and other defensive positions stretching in a semicircle from the Mystic River to east of Roxbury. His energy was tremendous; he could constantly be seen riding round the defences, and often with him were two subordinate major generals – Horatio Gates and Charles Lee. Both of these men had seen service in the British Army, both had settled in Virginia, and both were ambitious, but there the similarities ended. Gates, who was soon to be sent to Canada, and would later (with considerable help from Benedict Arnold) defeat Burgoyne at Saratoga in October 1777, was a kindly man and a good administrator, but in the field unimaginative and over-cautious; Lee on the other hand was a dominant and dynamic personality, arrogant, opinionated and charmless. His experiences in both the British and Polish armies gave him the professional edge over many of his senior colleagues, something of which he was never slow to remind them.

During the time that Washington was busily reorganising his rabble of an army and moulding it into an efficient fighting force, the British remained supine in the Boston and Charlestown peninsulas. Conditions at first were not too bad, there was plenty of food and accommodation was adequate – even luxurious for the senior officers. Boredom, an inevitable accompaniment of all the besieged, was the principal complaint, and Gage did very little to relieve it. Beyond improving the defences, and in particular those on Bunker Hill, and sending out the very occasional sortie to test the enemy strength, he was content to sit back and await promised reinforcements. He must have known, for his intelligence system was fairly good, of the many difficulties that faced Washington in command of an army that in numbers may have been superior to his own, but whose troops were in no way equipped to resist a determined offensive. And yet he made no attempt to break up the encirclement, nor did he see the need to occupy Dorchester Heights, which was a grave error, for as soon as Washington had sufficient cannon he would not leave this most important feature untenanted.

Gage was recalled in September, and left Boston on 10 October 1775. His replacement by Howe had been under consideration for some time, and cannot have been influenced by an interesting letter of 20 August written by Burgoyne from Boston to Lord George Germain. The latter did not in fact succeed Lord Dartmouth as Secretary of State for the Colonies and so in charge of the American Department until early November, but perhaps Burgoyne had knowledge of his impending move. The letter was a general review of the American situation, and some passages are worth quoting. 'General Gage is an officer totally unsuited for the command, and to this many of the misfortunes the King's arms have suffered may be traced.' He then turns his attention to Admiral Graves. 'It may perhaps be asked in England, what is the Admiral doing? I wish I was able to answer that question satisfactoriy. But I can only say what he is *not* doing. That he is not supplying the troops with sheep and oxen, the dinners of the best of us bear meagre testimony, the want of broth in the hospitals bears a more melancholy one. He is not defending his own flocks and herds, for the enemy has repeatedly and in the most insulting manner plundered his own appropriated islands. He is not defending the other islands in the harbour, for the enemy landed in force, burned the lighthouse at noon day, and killed and took a party of marines almost under the guns of two or three men of war.'[10]

Burgoyne's reports throw interesting sidelights on the campaign such as, in this instance, the supply base for Boston. No less interesting, and very often sound (even if they sometimes underly his own ambitions) are his suggestions for winning the war, or for improving the immediate position. On this occasion he put forward a plan for relieving the situation in Boston by dividing the force (which in August he says numbers only about 5,200 British troops) and sending 2,000 to threaten amphibious landings along the New England coast, which was likely to compel the enemy to reduce numbers investing Boston, thereby giving the British an opportunity to break out. The amphibious force could then sail south for Long Island and New York.[11] It might have worked, but Howe was not prepared to try it, for he always disliked the idea of splitting his force. When this limited scheme was first proposed, Congress was still openly loyal to the King, and Burgoyne himself believed a peaceful settlement was possible; but in the event of the present twilight peace being eclipsed by all-out war, he also had ready a much larger, and all-embracing plan for victory.

It must not be thought that Borgoyne was the only self-imposed strategist in Boston with plans to win the war. Clinton, too, was busily engaged in offering criticism and advice. Like Burgoyne he had absolutely no faith in Gage, and he was particularly depressed at the chaos and confusion in the army immediately after Bunker Hill, which he attributed mostly to Gage. He took a longer view, and advocated sending a large force to Canada and New York, and maintaining communication between them through holding the line of the Hudson. This, and further thoughts on the feasibility of a two-pronged offensive from north and south, he was to put in letters to General Harvey, the

Adjutant General, and the Duke of Newcastle on 15 November 1775.[12] In essence his was a plan somewhat similar to that being devised in the minds of Burgoyne, Howe and, later, Germain; but Clinton, aware of the shortage of troops, had an alternative based on strong garrisons in the north and south, possession of the Hudson and its passes, and leaving the rebellion to exhaust itself with the help of naval activity against the Atlantic ports.

The fuzzy prolixity of these various memoranda would in time be refined and, in part, translated into action, but meanwhile when Howe assumed command in Boston there was no noticeable increase in military activity. He shared Burgoyne's view that there could still be a peaceful solution, and maybe he was not anxious to stir the pot without more troops, particularly as life in Boston that autumn for some was by no means unpleasant. There were parties and plays (written and produced by Burgoyne until he sailed for home on 5 December) in the imposing Faneuil Hall, and a cosy mistress for Howe. Mrs Loring, the wife of a Loyalist, was to follow the drum with Howe for the next three years; it is nice to know that her husband was suitably rewarded with lucrative employment for such unselfish devotion to duty.

However, none of these distractions kept Howe from carrying out the necessary duties of a garrison commander. He continued Gage's work of improving the defences and, like that general, he maintained strict discipline with plenty of use of the lash. They were certainly tough in those days; Private MacMahon of the 43rd and his wife Isabella were found guilty of knowingly receiving stolen goods, for which he was given (and apparently survived) a thousand lashes from the 'cat' on his bare back, and Isabella received a hundred on hers. They were then imprisoned for three months.[13] Burgoyne was not in favour of flogging, being ahead of his time in maintaining that discipline of the right sort could be obtained by other means, but the lash was to survive for many more years. When not being flogged for ill-discipline, the troops were sent on occasional sorties (one as far as Lechmere Point, which accomplished very little), but no attempt was made to seize Dorchester Heights. Howe had less excuse than Gage over this, for the latter with his commitment to the Boston Neck lines might possibly have had insufficient troops to hold the ground. But in September, Howe received substantial reinforcements of five battalions, and a further five in December 1775. (The 17th, 27th, 28th, 46th and 55th in September, and the 15th, 37th, 53rd, 54th and 57th in December. The 16th Light Dragoons were under orders to sail and General Fraser's Highlanders (the 71st) were raised and sent out at about this time.)[14] This failure to occupy Dorchester Heights was to result in an over hasty evacuation of Boston.

The government in London and the generals in Boston were now certain that New York would be a better place for the army, but Howe consistently refused ministerial promptings because he had not sufficient transport, and to divide his force could be dangerous. In this he was very probably right. He also had to consider the Loyalists who should not be left behind, and anyway he had raised and armed three useful companies among them. Two of these have already been mentioned (The Royal Fencible Americans, commanded by Colonel Gorham, and the Loyal Irish Volunteers under James Forrest), and the third was the Loyal American Associators whose commander, Timothy Ruggles, was a brigadier general.[15]

Meanwhile, back in England the news of Lexington (which reached London via American sources on 29 May, two weeks ahead of Gage's dispatch), and later that of Bunker Hill, sparked off a new series of long and wearisome disputes and disagreements in political and military circles. Gage, perhaps inevitably but somewhat unfairly, was apportioned most of the blame. It was, however, now seen that his request for 20,000 troops was not unrealistic, and immediate steps were taken by the government to increase the army, and – in the autumn of 1775 – by the King to hire the mercenaries already detailed. Even so General Harvey, in the absence of a commander-in-chief the senior military spokesman in England, roundly declared that there would not be sufficient troops available to conquer America by land, and he advocated a vigorous naval blockade, which had its supporters and possibilities.[16]

The door to a peaceful solution was still ajar, and the hopes of some were raised when on 13 August Richard Penn, a former governor of Pennsylvania, arrived in England as the agent of Congress bearing a petition for the King to consider means for peace and union. But his chances of success were virtually nil, for if there were to be any negotiations, King and Cabinet were determined that they should be with individual colonies, and not with an illegal Congress. Nevertheless, Penn might have expected greater courtesy than he received. No minister showed any interest in his mission, and it was not until he had been in the country for eight days that Lord Dartmouth (the Secretary of State) agreed to a meeting. The King, who was now in full control and set firm on repressive measures, war or no war, flatly refused to see him.[17]

Parliament met at the end of October, and very soon the Opposition was in full cry. Furthermore, it had the support of many in the country, for all those involved in commerce had much to lose in a war with America. Rockingham, Fox, Shelburne, Camden, Burke, Barré and the controversial John Wilkes (recently Lord Mayor of London) all spoke vehemently against the decision to wage war with its concomitant need to increase the army through the Militia Bills and to employ mercenaries. Burke brought in a bill containing clauses to end the dispute through renunciation of the right to tax, and the repeal of certain acts to which the colonists had taken great exception. It was a promising bill which even at this late hour just might have averted further trouble, but it was defeated by 100 votes.[18] In this same session, the government introduced their Prohibitory Bill,[19] which conctained further repressive measures and, as a belated and futile peace overture, the appointment of two commissioners empowered to offer pardons and discuss conciliation with Congress. In due course, Admiral Lord Howe and his brother William, as the naval and land commanders, were chosen as commissioners.

Ten days before the introduction of this bill, the ardent Lord George Germain had replaced the more leisurely Lord Dartmouth as Secretary of State for the Colonies. This was a post established as recently as 1768, and became additional to the existing two Secretaries of State, those for the Northern and Southern Departments who were responsible respectively for foreign affairs in northern and southern Europe. At this time the Secretaries of State were more important instruments of government than the First Lord of the Treasury, for the duties of prime minister were nebulous and had yet to be clarified. Unless the latter was a strong personality, which Lord North was not, there was frequently a lack of co-ordination in Cabinet and hardly ever collective responsibility. The First Lord of the Admiralty and the Secretary-at-War were nominally subordinate to the Secretaries of State, but the importance of the Admiralty in the American War gave Lord Sandwich as First Lord, virtually equal status to Lord George Germain.

Germain was a highly intelligent man of action and ambition, but with a somewhat surly disposition, and inclined to arrogance. He had considerable military experience having risen to the rank of Lieutenant General, but in 1759 at Minden he had failed to obey (not through lack of courage) an order from his superior officer to advance, and as a result of a court martial was declared 'unfit to serve his Majesty in any military capacity whatsoever.' His enemies, and he had quite a few, would sometimes (and unfairly) use this long past and contentious setback in his career as his unfitness to employ his military knowledge in shaping the strategical operations of the war. It was not that he lacked the knowledge, but that he could not apply it as frequently as he wished from such a vast distance with such hopeless communications. His interference with the generals has sometimes been exaggerated, nevertheless it was fairly considerable. He lost no opportunity to put forward his own ideas and to express his recommendations, and in the southern campaign he made the grave mistake of issuing these to subordinate officers over the head of the Commander-in-Chief. His relationship with his fellow ministers was outwardly amicable, but he could seldom work in harmony with any colleague, and he was inclined to shift the blame on to other shoulders when things went wrong.

Germain's appointment was sharply criticised, not least by Howe's Whig friends in the House, for it was well known that there existed a mutual antipathy between the minister and the Howe brothers, which did not augur well for the future conduct of the campaign on land or sea. Nevertheless it was not a bad appointment, and many worse decisions had been made in the year 1775, which was now almost over. There had been too many misunderstandings, miscalculations and mistakes by ministers, and too many missed opportunities by senior officers in a year that had brought the country to the brink of a major war.

As winter approached there was more pressure on Howe to abandon Boston, and the government suggested Long Island as a stepping stone to New York, where the Loyalist presence was greater, and where he could be supplied by sea just as well. But Howe still maintained he had not sufficient transport, and his troops settled down to endure an unpleasant winter. Food, clothing and fuel (although the latter was partially alleviated through the destruction of old wooden houses) were now becoming a grave problem. Supplies of food from the local islands were drying up, but the government dispatched huge quantities at great expense – 5,000 live oxen, 14,000 sheep, 10,000 butts of beer, large quantities of vegetables and coal. These precious cargoes helped, but would not last the winter, and anyway there were losses at sea, for winter storms were a serious hazard, and as initially the transports were not protected, American war ships made one or two spectacular captures.

By the turn of the year, conditions in Boston for the soldiery had become very bad. Cooped up in a small area – if they showed their heads too prominently in the outposts they got a sniper's bullet – with little to do, insufficient to eat, and in ragged uniforms that failed to keep out the bitter winter weather it was not surprising that discipline and self-respect collapsed, despite the efforts of Howe and his officers. Among the poorer class of civilian inhabitant, matters were much worse, for they had even less to eat, and had to contend with a nasty outbreak of smallpox, which, for some lucky reason, did not greatly affect the troops, although they had their share of dysentery and other complaints. All in all, this was a miserable situation of inaction for soldiers who had been sent 3.000 miles to fight a war.

Washington too had his problems, the chief of which was a dwindling army. The approach of winter made short enlistment very attractive. Many of his men's time expired on 31 December, and they were mostly keen to go, while others drifted away without excuse. In many instances Washington was well rid of them, but recruiting replacements was not easy, and there was much inter-colony friction. Nevertheless, with characteristic energy and ability he virtually disbanded one army and, during the winter, replaced it with another: a fine feat at any time, but with almost no ammunition and a well equipped enemy watching it was a remarkable achievement. Of one thing he could be thankful, that the opposing generals had permitted him to survive.

However, in early 1776 his situation began to improve quite considerably. The many requests he made to Congress received their approval. An army of 23,372 officers and men was to be raised (a totally unrealistic figure never to be realised), with much longer terms of enlistment; 300,000 dollars were delivered to his camp, and at the very beginning of the year Colonel Henry Knox arrived with 43 cannon (among them one 24-pounder and eleven 18-pounders), 14 mortars and two howitzers, which had been dragged over the snows from Ticonderoga. Moreover, Washington was to have an increased fleet, for orders were given for the building of 13 warships each mounting 32 or 28 guns.[20] The time had surely come to go over to the offensive, and drive the British out of Boston.

By the end of February 1776, Washington's preparations were well advanced. Large quantities of fascines, gabions, twisted hay, entrenching tools and medical supplies had been collected in Roxbury, and brought forward in 300 carts towards Dorchester Flats. On the night of 1 March, those guns all along the American line that could range on Boston opened up a heavy bombardment, which was repeated on the nights of the 3rd and 4th. A certain amount of damage was done, but the principal object was to deflect attention from what was going on near Dorchester Heights. In this it was eminently successful. On a moonlit night, other precautions had also to be taken to protect the troops and long line of carts as they crossed exposed ground, and for this rows of hay bales were used. At nightfall on 4 March, General Thomas set off at the head of some 1,200 men and, crossing Dorchester Neck, brought them safely up to the twin hills of Dorchester Heights.

Once on the objective, the Patriots worked as hard as their colleagues had done almost a year ago on Breed's Hill and, like them, had completed some very stout defence works by dawn. It had not been possible in the hard ground to make use of entrenching tools, but the carts had brought sufficient materials for the construction of two miniature forts linked to smaller redoubts. Howe was staggered by the speed at which these strongpoints had been built. In a long letter to Dartmouth of 21 March 1776, written as his transports lay at anchor off Nantasket, he said that the Americans must have employed at least 12,000 men.[21] He had never included time and motion in his military studies. But both he and Admiral Shuldham (who had replaced Graves) realised that the 20 cannon the Americans had hauled up the two hills into a commanding position rendered Boston untenable, and moreover could seriously jeopardise any evacuation by sea,

Something had to be done, and done quickly. That night (5 March) Howe detailed Lord Percy to command 2,400 troops and sail for Castle Island, preparatory to an assault on the Heights. But a violent storm blew up that severely buffeted the ships at anchor making it almost impossible for them to put to sea, and quite impossible for the troops to be landed.[22] The operation was therefore cancelled. This was just as well, for the American position was almost impregnable. Any assault had to be made up a very steep gradient and the Patriots had assembled a number of barrels chained together and filled with heavy stones. These sent plunging down the hill, and backed up with accurate fire from the riflemen, would have caused confusion and many casualties. There were to be no further attempts on the heights. In his letter to Dartmouth, Howe was to write, 'I could promise myself little success by attacking them [the enemy] under all the disadvantages I had to encounter; wherefore I judged it most advisable to prepare for the evacuation of the town . . . .'.[23]

The unexpected storm may also have saved many American lives, for Congress had instructed Washington to prepare a number of batteaux and troops at Cambridge for an amphibious assault on the west side of Boston. Washington had demurred at this order, but the time was yet to come when his and not Congress' wishes were paramount in military matters. The attack was aborted by the winds before it got under way.

The detail of events leading up to Howe's evacuation of Boston are not entirely clear, and have received many interpretations. Obviously, Howe hung on as long as he could for the expected extra transports but, once the Americans occupied Nook's Hill, from where they could enfilade the British lines at Boston neck and seriously threaten the town, it was time for him to be gone. Thomas did not attempt the occupation of Nook's Hill until 9 March, and his men were then blown off it by British batteries, but on the night of the 15th they came back. With their by now customary speed and efficiency, they had made a sufficiently secure stronghold by the morning.

By this time Howe had already made preparations to evacuate the town, and was busy seeing how many people and stores he could cram into the 120 transports available. At some stage an arrangement was made, or tacitly agreed upon by both sides, that provided the Americans ceased pounding the town, Howe would do his best to ensure that the place was not burnt or unduly plundered. In this he was fairly successful. Various figures have been given for the number of people evacuated, but it would seem that the overcrowded transports held about 9,000 troops (a figure which probably included Loyalist soldiers), and at least 1,000 loyal civilians. In order to accommodate the latter, perhaps as many as 200 cannon had to be abandoned, spiked and damaged but quite repairable, and large quantities of powder, shot and other stores were left for the Americans.[24] There seems to be no doubt about this, although Howe in his letter to Dartmouth says, 'Such military stores as could not be taken on board were destroyed.'[25]

Howe's transports went only as far as the lower harbour on 17 March and dropped anchor off Nantasket, where they remained for 10 days organising a more seaworthy distribution of cargo. On the 27th, they sailed for Halifax in Nova Scotia, where his men were to linger into June with no orders, nothing to do, and living conditions as bad as those in Boston. Washington had been puzzled and a little troubled by the delay in departure, but the five hundred Patriots selected to march into Boston showed no anxiety and were jubilant. These men had had smallpox; the main body of troops waited until 20 March when the town was declared safe.

* * *

The whole Boston business was a muddle from which the British did not emerge with much credit. In the summer of 1775, the government had been anxious for the town to be evacuated, and they were right, for there was very little point, and precious little prestige, to be gained by hanging on to it. There were ports offering better strategic prospects, particularly New York and perhaps Newport (Rhode Island) or even farther south. All five generals – Gage, Howe, Clinton, Burgoyne and Percy – shared this view, but by November when Howe (who by then had succeeded Gage) received official authority to evacuate he judged it too late and too difficult to move. The troops had therefore to spend a miserable winter for the most part in appalling conditions.

Howe is often criticised for what seemed to be two serious mistakes during the investment of Boston. The first was his failure to attempt a break-out, or at least to harass the American lines by large-scale raids, and the second was his failure to capture the Dorchester Heights. He probably could have fought his way out, but to what purpose and to where? And as for raids, these were bound to result in casualties and, although possibly morale boosters, there would have been little point in them when evacuation of the town was planned for the spring.

In the matter of Dorchester Heights, however, Howe must have made a serious error. The British defences were good enough to ensure that any assault Washington might make would be repulsed, whereas possession of the Heights by American artillery virtually made not only Boston but, more importantly, the harbour, untenable. Clinton had pointed out to Gage back in June the desirability of occupying the position, but nothing was done either then or later. It is no excuse to say that the British position was already over extended, for Dorchester Heights and Nook's Hill were vital features and the Americans should never have been allowed to get them first.

Their occupation hastened the need for evacuation, and although Howe had assembled enough transports to avoid the two embarkations he had said would be necessary in November, many Loyalists had to be left to their most uncertain fate. Much military hardware was also abandoned, and this may have influenced the curious decision to go to Halifax, where it was known that conditions for the troops would be as bad as, or worse than they had been in Boston. In the letter to Dartmouth of 21 March, already quoted, Howe makes no specific reference to guns but, while admitting that New York would have been best, he felt that the condition of the troops in the crowded transports and the chaotic state of the stores 'effectually disable me from the exertion of this force in any offensive operations.'[26] Possibly, but was it not more likely that the landing would have been unopposed, as it was later on Long Island? Howe would seem to have thought otherwise, but on what grounds is not clear.

In and around New York there were many friends, and episcopalian friends at that – the religious connotation was important and never far from the surface in these early days of troubled uncertainty – and that place would appear to have been a far wiser choice. To go to Halifax meant a temporary military abandonment of the colonies at a crucial time when many senior Americans were beginning to think in terms of a struggle no longer for better conditions within the British Empire, but for independence. Loyalists badly needed the physical presence of British troops.

CHAPTER 4

Missed Opportunities

WHILE HOWE WAS still cooped up in Boston, a plan had been formulated to send an expedition to the South where the governors of North Carolina and Virginia had reported most enthusiastically, and over optimistically, on the number of Loyalists ready and willing to take up arms in support of the empire. Both these men had assured Lord Dartmouth that, with a little British encouragement, any acts of rebellion would be quickly suppressed, and the South pacified. Lord Dunmore, the impetuous Governor of Virginia, had already caused considerable trouble in trying to impose the firm policy he thought necessary, and had been forced to take refuge in a British frigate. Governor Josiah Martin of North Carolina was equally unpopular and he too was grateful for the hospitality of a British man-of-war.

Howe was strongly opposed to the dispersion of troops that an expedition to the South would involve. It would mean the postponement of the plan to occupy the line of the Hudson and isolate New England, and force him on the defensive for 1776.[1] Nevertheless, Dartmouth was attracted by this apparent opportunity to gain important results at small expense. Almost his last act before handing over to Germain was to sanction the dispatch of a squadron of 10 ships from Cork, under Commodore Sir Peter Parker, to Cape Fear, in North Carolina, convoying troops to be commanded by General Clinton.

Josiah Martin had slightly more reason for his misplaced optimism than Lord Dunmore, for he had in his colony that strong Loyalist contingent from the Highlands of Scotland. From among these he appointed Donald MacDonald a brigadier general and Donald McLeod a lieutenant colonel, at the head of a force of some 1,600 Loyalists, who were ordered to rendezvous at Cross Creek. Their immediate plan was to march down the right bank of the Cape Fear river to the sea. There they would be at hand to join forces with Parker's and Clinton's troops to help restore order in North Carolina. But rebellion there was well established and the Patriots had mustered two battalions of North Carolina Continentals, as well as a fair number of minute-men and local militia. At that time, those Continentals were a ragged bunch of men, but they knew how to fight.

MacDonald began to march his Loyalists to the coast on 20 February 1776, but soon found his way blocked at Rockfish Creek by Colonel Moore's

Continentals. He therefore crossed the river and struck east towards Colonel Richard Caswell's assembled militia, who promptly fell back from Corbett's Ferry on the Black River to Widow Moore's Creek, where they took up a strong position guarding both banks of the only bridge that spanned the 30-foot creek. After detaching a small force to support Caswell, Moore marched the rest of his Continentals in the direction of Wilmington, which would have posed a threat to the Loyalists had they managed to cross the bridge.

The Loyalists were now commanded by McLeod, for MacDonald had been taken ill, and they found themselves in an awkward situation. McLeod could not know that Parker's fleet had been hopelessly delayed and felt it was his duty to be at hand when it arrived. By the night of 26 February, Caswell had withdrawn his Patriots from west of the Creek and partially destroyed the bridge; he had almost 1,000 soldiers and some artillery behind hastily improvised, but fairly effective, earthworks. In contrast, many of McLeod's 1,600 men had only knives or daggers, the Scots relying mainly on the broadsword: to withdraw would be difficult, but still possible. To storm the broken bridge across a difficult swampy approach would be suicidal. Nevertheless, impetuous counsels prevailed and assault they did, on the morning of 27 February. The action was all over in a few minutes. Although casualties at the bridge were under a hundred, more than half the Loyalist force and a good quantity of weapons were captured in the subsequent round-up.

This defeat took place over two months before Parker arrived off Cape Fear, and three months before he and Clinton decided to attack Charleston. But so bad were communications, that the dispatch which the Cabinet sent on learning of the disaster suffered by the Loyalists, and which advised abandonment of the operation, did not arrive in time.

Parker's departure from Cork had been considerably delayed through weather, and it would seem that Germain was never entirely happy about the expedition. In a letter to Clinton of 3 March, he said it was the King's command that he did not expose his troops to great loss or jeopardise operations in the north and that he was to rejoin General Howe at once if, after the arrival of the whole force at Cape Fear, he was of the opinion that nothing of advantage could be effected.[2]

When Parker did eventually sail, he had a very long and unpleasant voyage, and he was not off Cape Fear until 3 May, after 80 days at sea. He had more than 2,000 men with him and Major General Lord Cornwallis.[3] Meanwhile, Clinton had left Boston in December 1775 with a handful of troops, and sailing via New York and Virginia, where he spent some time with the Governor, he had been off the Carolina coast awaiting Parker for quite a while. There was then a further delay for later transports to arrive, and for the two commanders to decide what to do. Clinton had been in favour of sailing up Chesapeake Bay (apparently with no particular object in view) until the Governor of South Carolina, Lord William Campbell, persuaded him to attack Charleston, and this was agreed with Parker.

The immediate target was to be Sullivan's Island, which commanded Charleston harbour. The plan in outline was for the troops to be landed on Long Island (now Isle of Palms) from where it was mistakenly assumed that they could wade across to Sullivan's Island at low tide, while the fleet bombarded the key point on the island which was a still incomplete, but very solid, fort, later to be called Fort Moultrie after its gallant defender. Since the previous November, the Americans had been busy fortifying prominent points of Charleston, which lies between the Ashley and Cooper rivers, and also its outlying islands. In January Colonel Moultrie had been put in charge of Sullivan's Island and its defences. Nevertheless, by 4 June, when Parker's fleet appeared beyond the harbour bar, there was still much work to be done, and troops were thin on the ground.

Had immediate action been taken, Charleston might have fallen, but there was to be one of those interminable delays (only partly caused through weather) that characterised so many of the British operations in this war, and with the immediate arrival of General Lee from the north, new spirit was put into the Americans. Moultrie wrote in a letter of 4 June to the President of Congress that 'it was thought by many that his [Lee's] coming among us was equal to a reinforcement of 1,000 men, and I believe it was, because he taught us to think lightly of the enemy, and gave a spur to all our actions.'[4] This was generous of him, for Lee did not approve of the island as a suitable base, or of Moultrie as its commander. It was not until 10.30am on 28 June, that Parker signalled to Clinton, who with his men had been on Long Island for some three weeks, that he was underway.

Thunderbomb, a mortar-carrying vessel, led the fleet into the harbour and flung shells with some accuracy. However, most of them sank quite harmlessly into the island's extensive bog, while the thick palmetto logs of the fort withstood even the heaviest cannon shot. *Active* followed *Thunderbomb*, and then came Parker's flagship *Bristol* with Lord William on board, with *Experiment* and *Solebay* also in the squadron's first division. These warships, mounting from 28 to 50 guns, stood off about 350 yards from the fort, and for eleven hours pounded the island without much effect. Severe damage to the fort might have been done by two smaller vessels sent up the channel to attack its west side, which was still unfinished, but they were badly piloted and ran aground on shoals.

At midday, Moultrie's batteries, which had replied with spirit and with some success, ceased firing and Parker became impatient for Clinton to attack, for he felt the island could be taken. But Moultrie was only short of ammunition, and Lee, who had crossed to the island in the heat of the battle, together with South Carolina's President, John Rutledge, soon had the guns resupplied.

Meanwhile, when Clinton prepared to support the naval bombardment by attacking the north of the island, he found wind and tide had banked the water in Breach Inlet to seven feet, and he had to attempt the landing in a number of small boats. For this he had the help of a sloop, but the American force under

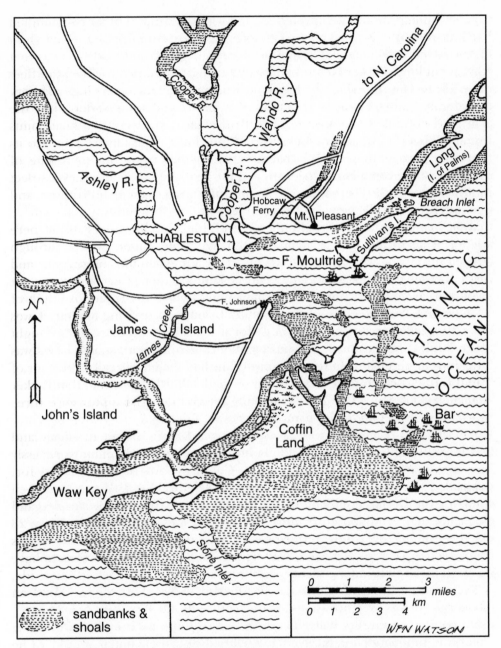

The Abortive Attack by Clinton and Parker on Charleston

Colonel Thompson was well protected, and with the support of an 18-pounder had little difficulty in breaking up Clinton's efforts to land. He therefore called off the attack, and contented himself with an ineffectual cannonade.

At about 9pm, Parker had to admit failure, and he ordered the fleet to withdraw. He had lost one ship, the frigate *Acteon*, almost all the others had suffered considerable damage (his flagship more than most), and it was obvious that the army was not able to co-operate. The British casualties were 205 killed and wounded, but the Americans had suffered little loss.[5] After lingering for some time in the harbour, hoping for a chance to redeem their failure, the commanders decided that the condition of the ships and the shortage of rations necessitated a return to New York. There was to be an unhappy postscript in the recriminations that lasted for a while between the two commanders. Clinton was determined that the blame for failure should not rest with him, and he at once dispatched a letter of justification to Germain. Eventually, Parker intimated that he had advised the attack, which may have been a true admission but a generous one, for it had all the appearances of being a joint decision. Germain, too, was anxious to avoid any blame. In a Parliamentary debate at the end of October 1776 he said, 'Sir Peter Parker's expedition failed from arriving too late: I am not responsible for its lack of success, for it was planned before I came into office.'[6] But if he had really felt uneasy about the operation, as well he might have done on receiving Howe's disapproval, he had the time to call it off.

On 4 July, before Parker and Clinton had left Charleston harbour, the American Continental Congress, issued the Declaration of Independence. The way had been prepared by a pamphlet entitled *Common Sense* written by Tom Paine, who had recently arrived as a settler from Britain. It had an immediate and immense appeal, almost certainly influencing South Carolina to declare her own independence in advance of Congress, and it must also have persuaded the leading men in Congress in their decision that the time had come for the Declaration. The result was Thomas Jefferson's masterpiece of eloquence and erudition enshrining the principles of a Christian civilisation. On the evening of 4 July Jefferson's Declaration (with some of the more controversial passages removed) was signed, and the American colonies had declared themselves independent communities.

There was much rejoicing throughout America; in New York the gilded statue of George III was overthrown and in due course its 4,000 pounds of lead were used to produce musket balls. There were also some parts of Britain where American Independence received unqualified approval, but it should have been obvious that the new country could not stand unaided, and that overtures would be made to France and Spain, which countries were sure to lend a willing ear. This ought to have given added urgency to the British government, and to its generals, to bring the rebellion to some sort of conclusion. But as will be seen, the generals, anyway, did not appear to get the message.

At the time that these momentous events were taking place in Congress, General Howe was assembling troops on Staten Island preparatory to an assault on New York. He had arrived from Halifax with his original Boston force on 2 July, and 10 days later he was joined by his brother, Admiral the Viscount Howe, in command of 52 ships of the line, 27 smaller vessels and 400 transports.[7] The Admiral was an experienced and skilful sailor and, like his brother, a very courageous officer. Also like his brother, he was a staunch Whig, who had never favoured the war, and he became very conscientious in attempting to fulfil the role of peacemaker with which the government had saddled him. In the course of the forthcoming campaign, which lasted until the end of the year, both brothers – but particularly William – were to be severely censured for a dilatoriness attributed to their desire to spare Washington and his men annihilation. (Lord Howe's role as commissioner got off to a bad start with the return of his first two letters to Washington for the somewhat puerile reason that they were addressed to the general as George Washington Esq. It was quite usual at the time for senior officers to be addressed as Esquire, and Washington must have known this.)

Germain had been sending massive reinforcements to Howe. These arrived during July and early August and, in the latter month, Howe's army was further increased by the arrival of Clinton and Cornwallis from the South with their 2,000 soldiers. Almost every source gives a different total for the number of men Howe had under command by the middle of August (and that applies to Washington's army as well), but it was around 30,000 of which a quarter were foreigners. Howe states in his Narrative that on 22 August he put ashore at Gravesend Bay, Long Island, between 15,000 and 16,000 troops.[8] The Americans had about 9,000 men on the island. The delay of over a month in making this move has often been criticised, but time was needed to organise the army into seven newly formed brigades, and Howe was short of important equipment until the middle of August.

The American force was under the overall command of the veteran General Putnam, who had under him Generals Lord Stirling and Sullivan (William Alexander claimed the Earldom of Stirling. His claim was rejected by the House of Lords, nevertheless he insisted on using the title.) Beyond their main position on Brooklyn Heights there were a series of wooded hills that formed a ridge crossing the island from north-east to south-west, and Putnam sent about 3,000 men under Sullivan to occupy this ridge. The British on landing took position along a line from Gravesend Bay to a little beyond the village of Flatland, commanding a number of roads that led directly to the American advanced position.

Howe's tactical plan, and the victory that crowned it, was a splendid example of the art of war even though accomplished against an enemy that was numerically greatly inferior.* Using the navy to operate against the American right and watch New York, Howe directed the army in a three-pronged attack.

---

*It is likely that Clinton, with his knowledge of the ground, had a good deal to do with the making of the plan.

General Grant was to take a force along the coast road against the American right under Stirling, von Heister's Hessians were to put pressure on the enemy's centre, while Howe, Clinton, Cornwallis and Percy took the bulk of the army in a wide right hook to come in on the left and rear of Sullivan's line of hills.

The action began in the early hours of 27 August, and the three manoeuvres were executed with perfect precision. Grant's men had the heaviest fighting, and it was more than four hours before they drove the Maryland and Delaware battalions back with heavy losses onto Brooklyn Heights. Howe's flanking force on the American left gained a large measure of surprise, and quickly rolled up that flank, taking a huge number of prisoners including General Sullivan, while the Hessians in the centre made good use of the bayonet. The British casualties were a litte under 400, while those of the Americans have been variously estimated as between 1,000 and 2,000, probably about 1,200 most of whom were prisoners.[9]

Howe was urged by his subordinate generals to go forward against the Heights while the Americans were still off balance, but, as at Bunker Hill, he showed a reluctance to bring the matter to an immediate conclusion. He said, 'The only advantage we should have gained would have been the destruction of a few more men, for the retrat of the greatest part would have been secured by the works constructed upon the heights of Brooklyn.'[10] It was also a fact that during the battle Washington had reinforced this strong defensive site with 3,500 fresh troops. And so Howe was surely right with that decision. However, it is debatable whether he was right to take no action in the next twenty-four hours and allow the enemy, helped by a sea fog, to evacuate all his troops to Manhattan Island under the very noses of the British sentinels. He still had the advantage of numbers, and the rain-soaked Americans, with dampened spirits and powder, were unlikely to have withstood the resolute use of bayonets. However, Howe was always mindful of the need to be sparing with his troops, for as he wrote later, 'even a victory, attended by a heavy loss of men on our part, would have given a fatal check to the progress of the war, and might have proved irreparable'.[11]

Lord Howe said his ships could not beat up against the strong north-east wind and enter the East River to interfere with the American withdrawal, or prevent Washington from evacuating troops and stores from Governor's Island. But there were those who said he could have come up on the flood tide and anchored. Be that as it may, the weather did play an important part in operations throughout the war to an extent that is not always recognised. However, on this occasion it seems more probable that it was not the weather but the fact that neither Howe brother was in the least interested in taking this excellent opportunity of smashing Washington's entire force, and thereby possibly ending the rebellion.

It is impossible to be certain what were the true feelings of Richard and William Howe in this matter of conquest or conciliation either on this occasion or (in William's case) later in New Jersey. Historians have tried for many years

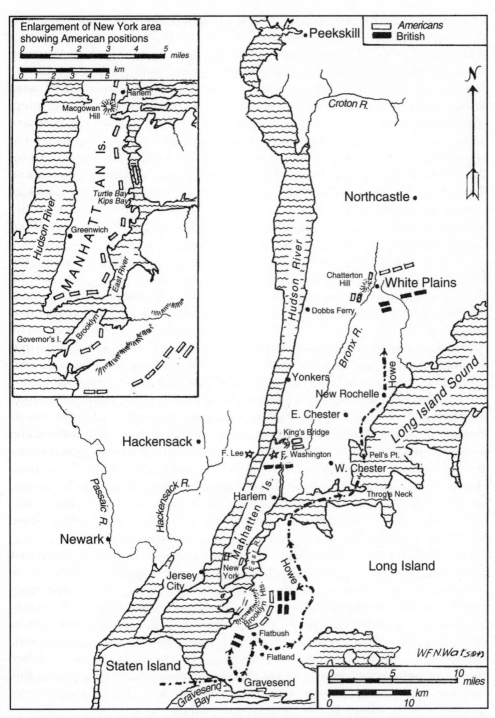

Enlargement of New York area showing American positions

Americans
British

Peekskill

Croton R.

Northcastle

Harlem

Macgowan Hill

Turtle Bay
Kips Bay

Greenwich

Chatterton Hill

White Plains

Dobbs Ferry

Yonkers

New Rochelle

E. Chester

King's Bridge

Hackensack

F. Lee

F. Washington

Pell's Pt.

Harlem

W. Chester

Throg's Neck

Newark

Jersey City

New York

Long Island

Staten Island

Flatbush

Flatland

Gravesend

WFN.Watson

The Actions at Long Island, Fort Washington and White Plains, 1776 to interpret their writings and sayings on the subject, and have come to different conclusions. The consensus is that the brothers were never heart and soul in the fight, and at this time were anxious for a peaceful settlement. But there are those ready to swear that their primary motive was to crush the rebellion by whatever means, and that Washington's army escaped to Manhattan and avoided destruction through military misunderstandings. We have to choose between what the brothers, and others who were present, wrote and said, and what were the actual facts. In every such case it is usually better to rely on facts rather than personal explanations and opinions written or spoken at a later date. The facts in this case do not indicate military incompetence beyond a reluctance to finish the job, which could have been accomplished without undue loss of life.

The Admiral, certainly, was more interested in peace, and had sent the captured Sullivan to Congress with further overtures. That assembly very reluctantly appointed a committee of three to meet Lord Howe on Staten Island. The whole business was an insulting farce, for it was soon apparent that the government had given Howe nothing to offer other than the granting of pardons, and British protection to those colonies ready to return to their allegiance.

This had wasted a good deal of time, all of value to Washington, and still General Howe tarried on Long Island until 15 September when, under cover of the fleet, now able to sail up both the East and Hudson rivers, he landed his force on Manhattan Island at Kips and Turtle Bays. Washington had decided to abandon New York, and had troops along the East River, but they mostly fled at the approach of the British. However, Putnam still commanded 4,000 men in New York who, it now seemed, would be trapped and destroyed by the British advance across the narrow island. Legend has it, although it seems almost unbelieveable, that these ragged, dispirited men were able to pass to safety up the Hudson River road, within a short distance of the British army, because that army's general was pursuing his love of dalliance at the richly laden lunch table of Mrs Robert Murray, in her house on what was later called Murray Hill. Whether or not that is true, Putnam's men escaped and, with the rest of the army, took up a strong position on Harlem Heights.

During the next three weeks Howe dug himself in across Macgowan's Hill, and made one or two probing attacks on the American entrenchments, in which both sides suffered some casualties with no significant results. It was not until 12 October that he put into practice the fairly obvious manoeuvre of sailing and marching round the American line, thereby cutting Washington off from his Connecticut supplies. Howe had three reasons for this further delay: he hoped the American army, which was certainly in very bad shape, might dissolve without the need for another major attack; he had the considerable inconvenience of a disastrous fire (almost certainly started by Patriots) that broke out in New York five days after he had taken the place; and, on 19 September, the brothers made yet another attempt to achieve a peaceful solution. This latter took the form of a joint declaration promising in the King's name certain concessions which went beyond those the government had stipulated, but still fell well short of what Congress demanded.[12]

When Howe stirred himself he could be a good general, which leads one to think that, apart from his habitual idleness, there must have been some underlying purpose in the frequent appalling lapses that allowed Washington to survive. His campaign from the disembarkation of his troops at Pell's Point all the way to Trenton had considerable merit and great possibilities, but there was no sense of urgency, and the one or two impressive tactical successes lost their value through Howe's almost obsessive desire to relinquish his prey.

Leaving Percy with three brigades on Macgowan's Hill, Howe sailed the rest of the army to East Chester. Here he proceeded to march north to New Rochelle where, on 23 October, he was joined by the recently landed second division of Hessians under General Knyphausen. The march continued at a fairly leisurely pace northwards, parallel to the Bronx River. Washington had opposed Howe's original landing at Throg's Neck, forcing him to re-embark for Pell's Point. The delay this involved enabled Washington to disengage from Harlem Heights and take up an extended position along the west bank of the Bronx, as well as to prepare a position at White Plains.

As soon as Washington understood Howe's purpose, he moved his troops to White Plains, detaching a strong force to occupy Fort Washington. This had checkmated Howe's intention of getting behind the Americans, and left him little option but to make a frontal assault. On 28 October, a mixed British and Hessian force attacked an outpost of Washington's line known as Chatterton Hill. They managed to drive the Americans off it, but at a loss of over 300 men killed and wounded. Howe then planned an attack on the main position for 31 October, but bad weather caused him to cancel it, and for the next two days he did nothing while awaiting reinforcements from Percy. But Washington knew his troops were not yet ready to face a major battle, and so he withdrew towards the Croton River, to a position at North Castle which was virtually impregnable.

Howe now acted with unwonted celerity. Leaving Washington at North Castle, he moved towards Fort Washington, with the intention of taking that fort and then, crossing the Hudson, occupying New Jersey as far as New Brunswick. This would pose a serious problem for Washington. Although he had secured his communications to the north-east, those to the west would be threatened by any move by Howe across the Hudson. Washington reacted equally swiftly by sending 3,000 men under General Heath to Peekskill to safeguard the Highlands, 5,000 under Putnam to cross the Hudson and make for Hackensack, and General Lee (recently summoned from the South) was left at North Castle with 7,000 men.[13] British ships had already found that the American batteries on the Hudson were not able to prevent their sailing up that river, and so Washington now had doubts on the value of holding the two forts (named Washington and Lee) on either side of the river. However, General Nathaniel Greene, who, after Washington, was American's finest commander, disagreed, and Congress supported his view. Washington therefore forbore to give Greene a positive order. This was most unfortunate, for Washington was right.

When Howe left White Plains, he sent a strong force of Hessians to hold King's Bridge at the north of Harlem Island, and marched the rest of his army to Dobbs Ferry and then south to confront Fort Washington. The Fort's garrison of over 2,000 men was commanded by Colonel Robert Magaw. As a building it was of little consequence, but it had a commanding position, standing above steep and rugged ground which cut through woodland, rocks and ravines. These natural obstacles had been strengthened by a number of well placed abatis, and both Greene and Magaw considered the fort impregnable. On 15 November, Howe's summons to surrender had inevitably been rejected, and General Knyphausen was ordered to take the fort by storm on the morning of the 16th.

The attack was carried out by four columns with strong supporting fire from two frigates, and batteries firing from redoubts that Howe had had constructed on the east bank of Harlem Creek. The hopelessly outnumbered and sternly embattled Americans made good use of ground and their reinforced outworks, and Howe's men had to fight for every yard of ground. Gradually the Americans were pushed back into the main building, but this was too small to hold the whole garrison, and in struggling to wield their arms they merely choked their art. At 1pm, completely surrounded, and faced by an overwhelming number of bristling bayonets, Magaw surrendered. A large arsenal of guns and ammunition was captured, and most of the garrison, for their battle casualties had not been great. The British had lost some 450 men. The Hessians had been well supported by the Guards, Highlanders and infantry of the line, nevertheless they suffered the bulk of the casualties, and it was perhaps their finest battle of the whole campaign. Fort Washington became (for a short time) Fort Knyphausen.[14]

It was known that Greene held Fort Lee with about 2,000 men, and so Cornwallis was now ordered to cross the Hudson at Yonkers and, at the head of a flying column of 4,500 men, make straight for the fort. So well and quickly was the operation executed that Greene and the garrison barely had time to escape. They left behind 12 men (hopelessly drunk), all the stores, tents, provisions and a good number of cannon. Howe now possessed all of Manhattan Island and much of the ground to the north; New Jersey and the road to Philadelphia (should he choose to take it) lay before him with seemingly nothing of consequence to bar his way.

Washington's fortunes were now at their nadir. He had crossed the Hudson on 14 November and witnessed the fall of Fort Washington from Fort Lee. He had with him about 4,000 very dejected men, who were deserting with considerable frequency, and these he led towards Brunswick. To add to his troubles, he was faced with near insubordination from that unstable, embittered man Charles Lee. Washington had repeatedly urged him to leave

North Castle and bring his force across the river, but Lee took little notice, for he wanted independent command. When he did eventually move south, he managed to get himself captured when sleeping ahead of his troops in an unguarded tavern. Washington lamented his loss, but the Americans were well rid of him, for his loyalty to their cause had the consistency of a weathercock.

What followed Cornwallis's taking of Fort Lee has never been properly explained. Howe had instructed Cornwallis to pursue Washington as far as Brunswick, in which area Howe was contemplating winter quarters. Captain Ewald, who commanded a Jäger company in the campaign, noted in his diary that the march to the Raritan River was constantly contested by small parties of the enemy, and although these were quickly brushed aside they were sufficient to prevent contact with Washington's main body, which on reaching the river destroyed the bridge and continued towards the Delaware.[15] Cornwallis was at Brunswick on 1 December, where he was met by a number of distinguished Loyalists from Philadelphia who urged him to press on, for Washington's army, now down to 3,400 totally exhausted men, could be easily routed and Philadelphia would offer no resistance. However, Cornwallis would not agree to an immediate resumption of the chase. He said, quite correctly, that his orders were to go no farther than Brunswick,[16] and he also needed to look over his shoulder at Lee's army now known to be on the move.

Howe, with a fresh brigade, joined Cornwallis in Brunswick on 6 December, and that afternoon the advance continued.* At Princeton the British were scarcely an hour behind Washington's rearguard, but the day was far gone and Howe decided to go into cantonment. On the morning of 8 December the march was resumed, but by then Washington had gone beyond recall. On reaching the Delaware the Jägers were sent to seize the American rearguard at Falls Ferry, but Ewald wrote 'the last boats were already leaving the shore when we were still about three hundred paces away.' Attempts to find boats were useless, for Washington had taken care to remove them all. Howe was content to leave Washington to recoup his fortunes in Pennsylvania, while he himself made plans for winter quarters. On Cornwallis's suggestion, and in order to offer protection for the Loyalists, these were dangerously extended to as far south as Burlington.[18]

In reviewing the year 1776 Howe (now General Sir William Howe GCB, the King having thus rewarded him for his victory on Long Island) may have had some satisfaction, for the foundations had been laid for his grand strategy. New York had fallen, and on 1 December Clinton had sailed for Rhode Island where he had met no opposition, therefore the way up the Hudson, with the supporting movement from the east, now seemed perfectly possible. Moreover, as Canada (but not Ticonderoga) was clear of enemy an advance from north to south could be undertaken. Indeed there was a feeling in the British camp that

*Ewald says that Cornwallis had in fact decided to march before Howe arrived on the 6th.[17]

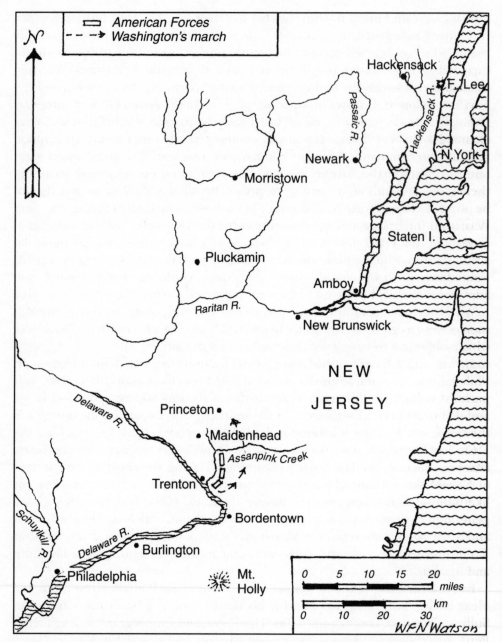

The Advance to the Raritan and Delaware Rivers

after a peaceful period in winter quarters the spring campaign would quickly finish the business off.

This was one side of the coin; on the reverse was the fact that Howe had let slip a major opportunity to win the war. The reasons why Washington's army was more than once allowed to escape, and in particular after the capture of Fort Lee, remain a mystery. Howe publicly and angrily denied that there was any 'private advantage'[19] involved, and this can be accepted. But a close examination of events inclines one to look for some reason other than a purely military one, for Howe was not a bad general, and Cornwallis was a very good one. It will not do for Howe to say he altered his original plan only on seeing there was a possibility of getting to Philadelphia; he should have altered it when he saw (if indeed he ever did see) there was a near certainty of routing Washington. Perhaps the kindest interpretation is Howe's habitual reluctance to engage in a rapid pursuit. But there must always remain the possibility of wishing to keep faith with his Whig cronies, or with his King. The Whigs never wanted the war, and George never intended annihilation, he merely wanted to administer a sharp lesson to bring an erring people back to their allegiance.

Whatever may have been the reason for these apparent lapses, Howe was soon given cause to regret it, for Washington possessed a tenacity of purpose which drove him from brave resolve to desperate execution. His retreat of some 90 miles in most difficult circumstances was a fine feat, and now in Philadelphia he did not despair. He bore his tribulations with courage and patriotism, and immediately set about pressing Congress for better conditions and terms of service for the new army he would organise in 1777. And before the old year was out he would execute a stroke of great daring.

On reaching the Delaware, Howe and Cornwallis returned to New York, the latter to prepare for his departure to England. He had handed over command of New Jersey to General Grant, who soon received information that there was likely to be an attack across the Delaware. Grant discounted this, for it was generally thought that Washington was in no condition to mount an offensive, and anyway the Delaware was icebound.[20] The 3,000 or so Hessians posted at Trenton and Bordentown were likewise unimpressed with this information, and settled down to make the best of the Christmas festival.

But the Delaware was not completely icebound, and Washington had sterner ideas about spending Christmas. With the arrival of Gates from the north, Sullivan (recently exchanged) with what remained of Lee's force, and the Pennsylvania militia, he had just under 6,000 men. For many of them their service time was due to expire at the end of the month, and they all needed a morale booster. Washington was determined to give it to them. He divided his army into three columns for crossing the river; on the right Colonel Cadwalader took Gates's place when the latter went off to Philadelphia, and he would be opposed to von Donop's men at Bordentown and Mount Holly; the centre column was under General Ewing, and Washington at the head of 2,400 men would cross the river a little way above Trenton. He and Ewing would attack Colonel Rall's 1,500 Hessians.

On Christmas night the moon was riding high, but the chilling wolfish voice of winter could be clearly heard in a gathering snow storm. Of the three Patriot columns only Washington's got across the river, Cadwalader and Ewing were defeated by large blocks of floating ice. Washington's party took 10 hours to cross, and then had an eight mile march to the village. His attack went in at two points at exactly 8am. Howe's account of the affair in his letter to Germain of 29 December says that Rall, being informed that the Americans had crossed the river, sent his own Hessians to support an advance piquet, and formed up the rest of his force in front of the village.[21] For some reason – allegedly the inclement weather, but possibly the previous night's celebrations – the morning patrols had been cancelled. Consequently, the piquet was surprised and, in falling back on Rall's regiment, caused considerable confusion. Rall then led his men in to the attack, and was killed at the head of them. Shortly afterwards the whole force surrendered.

Howe attributes the disaster to Rall not obeying the orders he sent him to erect redoubts.[22] He makes no mention of the generally accepted theory that the Hessians were in no condition to fight. Just over 100 Hessians were killed or wounded, and 916 captured, 412 escaped to Bordentown or Princeton. The Americans lost no men, but half a dozen were wounded including a future President, Lieutenant James Monroe. Washington recrossed the Delaware soon after the action, but when he heard that the left bank of the river was unoccupied, for von Donop had hastily moved his men to Princeton after the battle, he crossed again and occupied Trenton.

News of the *débâcle* at Trenton soon reached New York, where it caused considerable consternation. Cornwallis immediately cancelled his sailing, and instead made for Princeton, calling in troops to a total of almost 8,000, and Howe also left New York for Princeton. From that place, Cornwallis marched on Trenton, being nibbled at all the way by American skirmishers. Washington, meanwhile, had brought Cadwalader across the river and both forces took up a position on the east bank of Assanpink Creek (which runs into the Delaware east of Trenton) with the crossings well guarded. Cornwallis, coming up with his tired troops, could not force the position and settled down for the night, having sent for the regiments he had left at Maidenhead (present Lawrence-ville), with which he was confident he could defeat the Americans.

But Washington was not to be caught. Leaving his camp fires burning, and making use of a convenient mist, he slipped away in the night and marched for Princeton and farther north. At Princeton his troops ran into reinforcements under Colonel Mawhood on their way to Cornwallis, and in a 20 minute battle he inflicted almost 200 casualties on them and took 230 prisoners, but he lost a number of officers including General Mercer whom he could ill spare. Writing to the President of Congress from Pluckemin (a village between the Raritan and Morristown), on 5 January 1777, Washington said that with 600 or 800 fresh troops he felt he could have destroyed all the stores and magazines at Brunswick 'and put an end to the war.'[23] This was wishful thinking, but it shows the spirit of the man at the head of an army still dangerously weak and in grave danger. It also shows the reason for British panic, for the capture of Brunswick would have been a serious matter.

As it was, Washington headed for Morristown which he reached on 7 January, and took up a strong position on the Heights. Putnam came up from Philadelphia to occupy Princeton, which together with Bordentown and Trenton had been evacuated by the British, and other troops cleared Hackensack in the north, and also Elizabethtown which rendered Newark untenable. Here matters were allowed to rest for almost six months. In a fine display of generalship, and spectacular audacity, Washington had galvanised a dejected army and in less than three weeks won two battles with it; taken over 1,000 prisoners; scattered an army twice its size; reconquered all of New Jersey less Brunswick and Amboy; and saved the rebellion.

None of this reflects much credit on the senior British commanders, and none of it need have happened. Washington had now got Howe's measure, but the latter did not appear to be unduly troubled, and was to enjoy the winter in New York in the company of Mrs Loring. But Clinton was on his way home, and he was not likely to give the government a favourable impression of the Commander-in-Chief. The ministerial rejoicings that followed the news of Long Island and Manhattan had now turned a trifle sour, and the government's hopes for a swift conclusion to the war became only a distant dream. Lord George Germain is reported as saying, 'All our hopes were blasted by the unhappy affair at Trenton.'[24]

* * *

Clinton's and Parker's expedition to the south was from the very beginning a badly muddled political and military exercise. Its original intention was not so much conquest as a demonstration of force to encourage Loyalist enthusiasm and enrolment, based largely on over optimistic reports from the governors of the southern provinces. On the strength of these reports Dartmouth conceived the idea of possessing a southern port from which to carry out attacks on seacoast towns in support of the Loyalists, and on 22 October he sent instructions to Howe accordingly. Before any reply from Howe could be expected this had developed into the Cape Fear project. There were then endless delays (admittedly some unavoidable) in dispatching the force and, with the ever present difficulty of time and space, there was no hope of informing the Loyalists of this, and thus perhaps curbing their impetuosity and so avoiding the fiasco at Widow Moore's Creek.

Dartmouth's original idea became vitiated by the joint commanders, Parker and Clinton. The former favoured an altogether more complex operation with the taking of Sullivan's Island as a stepping stone to the capture of Charleston. Clinton, on the other hand, having paid courtesy calls on two of the governors as he sailed south, clearly saw the need for permanent occupation by regulars if loyalist effectiveness were to be sustained. Furthermore, he was certain there would be untold difficulties in maintaining such a force out on a limb in the south. He therefore favoured a similar sort of operation, but in the more strategically favourable area of the Chesapeake.[25] He had had authority from Germain to abort the operation if he thought it advisable, but in his Narrative he was to write that 'the Commodore [Parker] and general officers, whom I consulted on this occasion, agreeing with me in opinion that the object before us promised very great advantages and was likely to be accomplished without much delay, we determined to proceed upon it now we were so near the port.'[26] A great pity, for it was an attack that should never have taken place, and which was a damaging failure.

Perhaps the greatest puzzle to solve for this period of the war is the conduct of the Howe brothers, in particular that of Sir William. It is an indisputable fact that he had Washington at his mercy on Long Island, on Manhattan and in New Jersey. Yet he had let him and his army slip through his hands. It is easy to suppose that he and his brother were not interested in destroying the American army, but merely in dealing a sufficiently severe blow for Congress to pay proper respect to Lord Howe's forthcoming peace overtures. And yet there must be a doubt, for William set about his campaign on Long Island as though he meant to drive the American army straight into the East River. He had overwhelming superiority in numbers, and his brother had the ships to cut off any escape from the island. To destroy the force in front of him, which was far the greater part of Washington's whole army, would have had a much better effect on Congress than paltry half-measures. And Howe must surely have known it.

To assault Brooklyn Heights would have been folly indeed; to lay siege was a sensible option, although it did not need much intelligence to know that Washington would be out of the back door if at all possible. The chance to stop him doing so was not easy, but perfectly possible. As has been seen, excuses were made by both brothers and they were not very convincing. The likelihood is that Washington owed his escape to their irenic motives, but it just could have been to bad generalship. In either case it was a dreadful mistake.

Closely bound up with these military manoeuvres was the peace offer that Lord Howe had made a condition of his accepting the naval command. The concessions that the Howes were empowered to make merely emphasised the importance of a complete victory, for it was painfully obvious that while there was still life in the rebellion Congress would not agree to disband their troops, dismantle their forts, and renounce all revolutionary organisations in return for the granting of limited pardons, the removal of certain trade restrictions, and no colonial tax – provided the colonies contributed sufficient for their defence. Furthermore, the Howes had very little room for manoeuvre, because the King, whose heart was not in the business, had concluded his long instructions with the words 'And it is our Will and Pleasure that, in the Discussion of any Arrangements that may be brought forward into Negotiation in consequence of a Restoration of the public Peace, you do not pledge yourselves in any Act of Consent or Acquiescence, that may be construed to preclude our Royal Determination upon such Report as you shall make to Us, as aforesaid, by one of our principal Secretaries of State.'[27]

Such ridiculous overtures for peace without a crushing victory had no hope of success. Lord Howe had done his best with his hands tied, but he was soon to be told by Congress that there could be no negotiation unless based upon the recognition of independence. There was still time to change that reasoning by force, but the government at home and the generals in the field seemed to think otherwise, for all the way to White Plains and across the Hudson to the Delaware there were to be more half-measures and a sickening lack of urgency.

There remains to be considered, for this campaign, the approach to the Delaware and the disaster at Trenton. Cornwallis's failure to continue the pursuit beyond New Brunswick has often been condemned. In his evidence before Parliament in 1779 he made one or two interesting statements. At one point he says his troops 'had been constantly marching ever since their first entrance into the Jersey's', but Ewald's diary clearly shows this was not so, there were three days when the army did not move. And again, having said he understood it was the general's wish to stop at Brunswick he added 'but had I seen that I could have struck a material stroke, by moving forward, I certainly should have taken it upon me to have done it'.[28] He considered no advantage could be reaped from further pursuit, for at least a day would be lost in repairing the bridge over the Raritan.

The fact that on arrival at Brunswick Cornwallis's men were in no condition to go on without rest and food must be accepted, but it seems strange that he could see no advantage in moving forward after his troops had had a brief rest. It will be seen later that Cornwallis was a man who did not hesitate to disregard orders if he felt there were advantages in so doing, and contemporary accounts make it plain that on this occasion there were enormous advantages to be obtained. A few days later Howe arrived at Brunswick, and even then there was time to trap Washington before he crossed the Delaware, but so leisurely was the eventual pursuit that the Americans were able to escape.

From a study of diaries, letters and memoirs (at least some of which should contain the truth) it seems obvious that in the last six months of 1776 there was every opportunity to crush the rebellion and create a situation in which sensible negotiations could lead towards a satisfactory future. This was lost through a series of military and political errors, and with the turn of the year Washington was virtually to save the revolution.

Through the advance to the Delaware, the British had already become over extended, and Cornwallis made this worse by his quite understandable wish to protect Loyalists by a further extension of winter cantonments. Howe, somewhat reluctantly, agreed to this, for there were advantages, but Clinton pointed out the dangers of isolated surprise attacks. Entrusting the defence of the Trenton line to the Hessians was no mistake, for they were good soldiers, but General Grant, who had been put in overall command when Howe retired to New York and Cornwallis headed for home, appears not to have inspected the defences and Rall, the commander on the spot, had not carried out his orders to erect redoubts. There was the usual underestimation of the enemy's ability, the Hessians were surprised and rolled back, the British were defeated at Princeton, and Washington ended by recapturing almost all of New Jersey.

It is usually said that Burgoyne's surrender at Saratoga in 1777 was the turning point in the war, for after the French intervention which it heralded, British hopes for success entirely evaporated. But Trenton was a significant landmark, for the affair there and its immediate consequences seriously deflated British invincibility, and American morale was consequently boosted. At least at Saratoga there were to be some excuses, but for the conduct of the campaign from the Hudson to the Delaware it is difficult to find much to say in praise of generalship.

CHAPTER 5

Plans and Counter-plans for 1777

THE YEAR 1777 was a crisis year in the War of the American Revolution. Had the original strategy been fully implemented it is just possible, although unlikely, that the way would have been opened for ending the war either in that year or early in 1778. But miscalculation, misunderstanding, muddle, bureaucratic inefficiency, delay and divided command all played their parts in producing two virtually unrelated campaigns, one of which ended in a major disaster that made it almost certain the war could never be won.

Since Independence, the Americans had been receiving a gradually increasing amount of indirect aid – money, arms, gunpowder, trading facilities – from Holland, France and Spain. It required only an overt sign that the new republic was capable of fortitude and constancy on and off the battlefield for France at any rate, and quite possibly the others, to take the opportunity of giving active support in much desired revenge on Britain. Washington had done his best to give this sign in the weeks after Trenton, and before the year was out it would be unmistakably shown that what had happened in New Jersey was no shadow without substance.

The first and foremost British plan for the 1777 campaign centred on holding the line of the Hudson and isolating New England, the linchpin of the rebellion, from the rest of the colonies. It originated in April 1775 when Lord Dartmouth had put it forward to General Gage, then commanding in Boston.[1] Howe was to elaborate on it that autumn and Carleton, after the failure of his earlier expedition from the north, suggested that with an increase of 100 men per company (approximately 4,000) he should make another attempt, and this time send a supporting detachment to operate down the Mohawk Valley, and possibly the Connecticut River.[2] A little later Burgoyne submitted a similar plan to Carleton's in his *Thoughts for Conducting the War from the Side of Canada.*[3]

The principal defect of all these original plans, which envisaged the northern army being joined by Howe advancing from New York, was that it was quite impossible (even with an army three times the size that Britain could send) to isolate the New England states in this way. The terrain was such that small parties of local militia could always get through, and anyway the most important traffic was by sea in small boats, which was something the Royal Navy, even if it had the ships available to seal the many ports, would have

Proposed British Advances, 1777

found it almost impossible to stop. Ministers in London found it difficult to understand this and, as can be seen, the generals who developed Lord Dartmouth's original concept had reservations, and proposed a variety of alternative or additional ideas. All of these, in so far as the Hudson Valley was concerned, relied solely on the seizure and occupation of ground, with action against Washington's army limited to a holding role. To win the war, the Americans had to be decisively defeated in the field, and Washington was unlikely to allow his army to be ground between the upper and nether millstones. Wandering up and down the Hudson Valley was never going to accomplish anything.

In any event, to ensure success the plans eventually adopted had to be properly understood, agreed and acted upon by the five principals – Germain, Howe, Carleton, Burgoyne and Clinton. Unfortunately, not all of these important prerequisites were obtained. All five men had ideas as to how the war could be won, and although in most cases these did not differ greatly, the approach to the problem was somewhat complicated by strained personal relationships. Howe and Clinton, his second-in-command, seldom agreed on anything; the latter was in England for much of the planning stage, pushing his own views, and hoping for command of the northern army should Carleton, who by now had an independent command in Canada, not be given it. Germain's unhappy relationship with the Howe brothers went back to the days of Minden, and he was on the worst of terms with Carleton, which probably lost the latter the northern command. Burgoyne, at least, was unencumbered by discord but, no stranger to ambition, he waited in the wings, ever elegant and suave, seeking glory. The most important difference of opinion existed between Germain and Howe, and it centred round the number of troops required, and the best strategy for winning the war.

Howe's first plan was sent to Germain on 30 November 1776.[4] He presumed that, in the spring, Carleton would renew his previous attempt to move south down the Hudson, and could not be expected to reach Albany before September. Howe proposed two major offensives, each with 10,000 men. One, based upon Rhode Island, was to attempt to capture Boston, the other would move up the Hudson to join forces with Carleton at Albany, and thus sever New England from the other colonies. Eight thousand troops would be ear-marked to keep Washington from moving out of New Jersey to interfere with operations on the Hudson, and small forces of 2,000 and 5,000 men would be left to secure Rhode Island and New York respectively. Assuming the success of the Hudson enterprise, Howe envisaged a major offensive on Philadelphia in the autumn and a winter campaign in the southern states. For this plan he would require 35,000 troops. As he had 20,000 effectives, he asked Germain for 15,000 reinforcements.

It was a plan with plenty of merit, but when it arrived on Germain's desk he was considerably shaken by the number of men it required. This was understandable, for the military establishment total for 1777 allowed only

57,000 men for all colonial garrisons, and that was a paper figure far removed from fact.[5] Here was the fundamental difference between Germain and Howe, for the latter needed a constant expansion of British arms whereas Germain had to think in terms of winning the war with the limited resources available. In this instance he did some juggling with Howe's figures, and by underestimating the likely losses through casualties and sickness, he convinced himself that with 7,800 reinforcements Howe would have his army of 35,000.[6]

Meanwhile, on 20 December, before Germain could have received the first letter and long before any answer to it could have reached America, Howe sent a second dispatch. In this he switched the emphasis to Pennsylvania, for he felt 'the opinions of the people being much changed', that they would be confirmed in their loyalty by the capture of Philadelphia. He therefore intended to abandon the attack on Boston, at least until receiving substantial reinforcements, and to take 10,000 men to Philadelphia. Nearly the same numbers as before would remain in Rhode Island and New York, but only 3,000 men would be stationed on the lower Hudson, 'to facilitate in some degree the approach of the army from Canada'.[7] At almost the same time, Lord Howe was writing to Lord Sandwich for eight additional ships of the line.[8] Sandwich, like Germain, was desperately trying to stretch his resources and refused to send reinforcements from the Channel Fleet, but promised a limited number of two-deckers.

It is difficult to know what prompted Howe's new thinking; it is probable that he realised his earlier request for 15,000 troops could never be met, and therefore that this plan, which only required a total of 19,000 men, was more realistic. And, of course, between the dates of the first and second dispatches, events in New Jersey had given him great encouragement, for Trenton was still to come. But whatever its reason, this new plan was to jeopardise the chances of the northern army. Howe always believed that that army could look after itself. He did not seem to appreciate the harshness of the country and the logistical problems the northern army would have to face. Moreover, without his advancing force, the tough New Englanders would be free to devote their full attention to that army.

Howe's amended plan reached Germain on 23 February 1777. In his reply to the first dispatch, Germain had been non-committal, but this latest one met with both his and the King's approval. He informed Howe of this in a letter of 3 March, adding that the King greatly favoured a diversion on the coasts of New England, and he hoped the Howe brothers would give it serious consideration.[9] Neither brother favoured this, and Sir William was disappointed to learn that Germain could now send him only 2,900 reinforcements and not 7,800 as promised.

The next important communication from Howe to Germain was written on 2 April, which was before he received Germain's letter of 3 March, and in it he made yet another distribution of troops. The most significant alteration was that he no longer intended to leave 3,000 regulars 'to facilitate ... the approach of the army from Canada', but instead placed 3,000 Loyalists under Governor Tryon with no orders for them to act offensively. But, of equal importance, the dispatch announced his intention virtually to abandon New Jersey and take his army by sea for the attack on Philadelphia.[10] There was some reasoning in this, for his lines of communication were very long and needed a lot of troops to guard them, and the Delaware would be a formidable obstacle in an opposed crossing. However, it meant that the main British army would 'be out of touch with New York until Philadelphia was taken. The dispatch arrived in London on 8 May, and was approved by Germain on the 18th. In his reply he impressed upon Howe that whatever he might meditate would be 'executed in time for you to co-operate with the army ordered to proceed from Canada'.[11] This letter was not received until August, by which time Howe had sailed for the south.

On 28 February, Burgoyne, who had not yet been selected to command the northern army, submitted his 'Thoughts' to Germain. Basically these were along the lines of the other plans in that the Ticonderoga-Hudson line should be taken and held by a series of blockhouses, and the northern army should effect a junction with Howe at Albany, thereafter remaining on the lower Hudson to enable Howe to take the main army south. Burgoyne seemed to be aware of the problems and dangers of the land route, for he pointed out that water transport could be used for almost the whole way. He also suggested the need for a diversionary force along the Mohawk Valley, and the use of the Connecticut River as an alternative in certain circumstances. He stipulated the need for 8,000 men and, like others, he overestimated the numbers and value of the Loyalists.

As if all these plans were not enough, General Lee, in the hope of mitigating his offence of desertion from the British Army, turned traitor and gave the Howes his opinion as to how Washington could be defeated. His plan, like others, aimed at severing the New England states from the southern ones, but was based on the capture of Philadelphia rather than on operations to the north. He gave an over-optimistic assessment of the loyalist situation in Maryland and Pennsylvania, assuring the Howes from his personal knowledge that there would be a large loyalist uprising on the appearance of a British army. He therefore advocated sending 14,000 men to operate in New Jersey and to capture Philadelphia, while 4,000 should go by sea to Chesapeake Bay, and occupy Alexandria and Annapolis. This plan was never seriously considered, and indeed was kept under wraps for 80 years when the handwritten document was discovered in the Strachey archives endorsed by Henry Strachey (secretary to the Howes at the time) 'Mr Lee's plan, March 29, 1777'.[12]

The conduct of any war is ultimately in the hands of the country's government or ruling cabal which, quite rightly, likes either to originate or to be closely involved in the strategical planning for any major campaign. The outline given above of the various plans for the 1777 campaign, and their alterations, which at one time was hoped would win the war, clearly shows the extreme difficulties imposed by distance in the days of sailing ships and, in particular, over the maintenance of communication between ministers in London and British commanders abroad. This has to be borne in mind when censuring those responsible for the muddles and mistakes that were undoubtedly made at the time and in the early stages of the campaign.

By the middle of March, the government had decided how the campaign was to be conducted. Howe's plan for Philadelphia had been approved, Carleton was to hand over the northern army to Burgoyne after retaining 3,000 men for the defence of Canada and Clinton, who was still in England and whose ruffled feathers at not getting the northern command had been somewhat smoothed by an honour, agreed to resume his post as Howe's second-in-command that summer. Before he returned to America, he did his best to impress upon Germain that Howe could not be expected to complete his campaign in Pennsylvania in time to give assistance to Burgoyne. But Germain ignored this wise counsel from a general with experience of the country, and persisted in writing to Carleton to say that Burgoyne's principal objective was a junction with Howe's army.

In this letter, dated 26 March, Germain said 'I shall write to Sir William Howe by the first packet',[13] but in fact he never did. There are many versions of the muddle that occurred in the Minister's office. The most usually accepted is that when on the point of leaving for the country Germain was reminded by his Under-Secretary, William Knox, that Howe had not received any instructions, he did not delay his departure, but agreed that instructions could be drafted by the Deputy-Secretary in time to catch the next packet. But all the Deputy-Secretary did was to send a covering note with a copy of the letter to Carleton. And so Howe never received specific instructions, although he knew very well what was required of Burgoyne and himself. In fact, had any separate instructions been sent to Howe, they would merely have foreshadowed those sent later to complete his campaign in time to assist Burgoyne, because both Germain and, at this juncture, Burgoyne, who was still in London and privy to the planning, were convinced that the latter could reach Albany unaided.

The only person who was not happy at the outset of the campaign was Clinton, who thought the plan was wrong, that it would fail, and that he and his small force in New York would be swallowed up by Washington. At the time, his fears were derided, but he was not altogether wrong.

The results of all this planning will be seen in the chapters which follow. Howe's campaign was a partial success, Burgoyne's a total disaster, though little blame can be attached to Burgoyne. The principal responsibility for failure must rest with the thoughts and actions of Germain and Howe before Burgoyne ever crossed the Canadian border. Germain had the duty of getting the Government's and the King's acceptance for the plans that he and the military commanders had formulated; he was an intelligent man with a certain amount of active military experience and, although he had no first-hand knowledge of American conditions, he had the opportunity to be well informed. The ultimate responsibility for the plans must be his. However, having issued them, he needed complete co-operation from the generals. Howe fell short on this.

If both campaigns were to be brought to a successful and more or less simultaneous conclusion, it was necessary for Burgoyne's army to be capable of reaching Albany unaided, or to be supported by a sizeable force from Howe's army acting on the Lower Hudson. If the latter was likely to be necessary, then it was essential that Howe either left sufficient men in Manhattan with positive orders, or that his campaign in Pennsylvania should be completed in time for him to render this assistance. In the event, Burgoyne's army could not reach Albany unaided and, as Clinton had pointed out both to Germain and Howe, the latter could never have the time to take Philadelphia and assist Burgoyne. Germain is often criticised for not having reiterated in his correspondence with Howe the need for co-operation with Burgoyne. But Howe was well aware of the plan, and genuinely felt that Burgoyne could look after himself, except in the unlikely event of Washington marching north, in which case, he told Burgoyne, he would come after him.

Howe must have known that there were scarcely enough troops in America to conduct two major campaigns simultaneously and, for that reason, he kept pressing Germain for more. Once he changed the original plan, which was the best one, he showed little enthusiasm for the role of the Canada army. Always resenting his lack of troops, he was not prepared to jeopardise his own plan unless there was a real need, which he could not believe likely. He made this quite clear in a letter of 5 April to Carleton, in which he said he would not have the strength to detach troops at the beginning of the campaign to act upon the Hudson, and that Burgoyne must get to Albany on his own.[14] Burgoyne, for his part, showed no lack of confidence in the ability of his army to achieve the task given it. He was probably over optimistic as to the likely loyalist support he would receive, for this was a universal fault of all the British commanders, and he may have misjudged the value of the Indian contingent he was hoping to enrol.

* * *

Burgoyne's story will be told later, but it is sensible to anticipate the causes of his troubles, for they stemmed very largely from the planning defects that have been described above. Two men must share the principal blame for this, Lord George Germain and Sir William Howe. Of the two, Howe, by the very fact that he was the man on the spot with a comprehensive picture of the local situation, has perhaps the most to answer for. The genesis of the disaster at Saratoga lay in much muddled correspondence between London and New York, and the lack of any firm directive. Germain, who had the responsibility for directing the war, might have handled the Commander-in-Chief more firmly by insisting on the original plan if – as seems likely – he thought it best, And once Howe had intimated his preference for a second campaign, the approval given on 3 March should have *then* been made conditional on his executing the plan 'in time . . . to co-operate with the army ordered to proceed from Canada'. It did not need a genius in the study of time and space to realise that by 18 May (the date this instruction was sent) co-operation was virtually impossible.

Thereafter the gravamen against Germain must be a seeming indifference. There was an inexcusable muddle over certainly one dispatch; a failure to impress upon Howe the degree of urgency with regard to Burgoyne that he (Germain) felt necessary; and, at times, a sense of bewilderment, as shown by his letter of 24 June to William Knox, in which he wrote 'I cannot guess by Sir Wm. Howe's letter when he will begin his operations, or where he proposes to carry them on.'[15] This after all the earlier correspondence.

But it was Howe who held Burgoyne's fate in his hands. It is clear from his correspondence that he had no intention of being in a position to assist him, for in spite of what he was to say about going up the Delaware to be nearer if needed, he made no effort to hurry forward his campaign, which alone might have made this possible. No doubt he genuinely believed Burgoyne was all right but that, apart from showing a proper lack of appreciation of the difficulties which Burgoyne might meet, does not excuse his failure to make any provision in case things went wrong, as they often do in war. His failure to leave Clinton with definite orders to take action on the lower Hudson, and sufficient regular troops with which to do so, was a very serious blunder. Whether his adherance to the original plan would have brought an end to the war is extremely doubtful, but it would almost certainly have avoided a British catastrophe. The Philadelphia campaign could never have justified his revised planning unless it had been begun much earlier and achieved much more.

CHAPTER 6

The Taking of Philadelphia

WHAT MILITARY ACTIVITY there was in New Jersey during the winter of 1776–77, and indeed well into the spring, was almost entirely on the American side, and consisted of irritating hit-and-run raids by small numbers of troops against British encampments, foraging parties, and installations on the line of communication. The performance of Sir William Howe that summer amazed and bewildered both friend and foe, for the two props of his strategy – the defeat of Washington's army and occupation of territory – tended to become confused, and his thinking generally appeared to lack clarity and coherence.

There has been no lack of criticism for Howe's failure to attack Washington in his winter quarters, or for his lateness in starting the campaign that summer, which made it virtually impossible for him to capture Philadelphia in time to give aid to Burgoyne. It is easy to condemn Howe, for he was by nature somewhat indolent, and he certainly wasted a lot of time in those early months. But he was not nearly as incompetent as is sometimes made out, and when later he defended himself against these charges, his explanations were in some instances quite convincing, and the competent Cornwallis gave evidence much in his favour.[1]

Winter campaigning in the 18th century was not to be lightly undertaken, for the weather could play havoc with supply systems as well as the general health and condition of the troops. Washington's army may still have been in a deplorable state, and lacking much, but it was strongly positioned. Howe's principal objective was Philadelphia, 80 miles to the north. If he risked battle against the Morristown defences, he had to win, and this was by no means a certainty. He was well aware of this, and so why attempt a dangerous winter assault when – as he could be excused for thinking – time and numbers were still on his side? In refusing battle both at Morristown and Middlebrook, an even stronger position to which Washington had moved at the end of May, Howe was undoubtedly right.

Against the indictments of delay and seeming indecision, he is on less sure ground. In April, when he decided to go by sea to Philadelphia, he warned Germain that there would be a delay in opening the campaign[2], but for his almost total inactivity from then until early June, the word 'delay' seems decidedly an understatement. There was no apparent reason for this inactivity

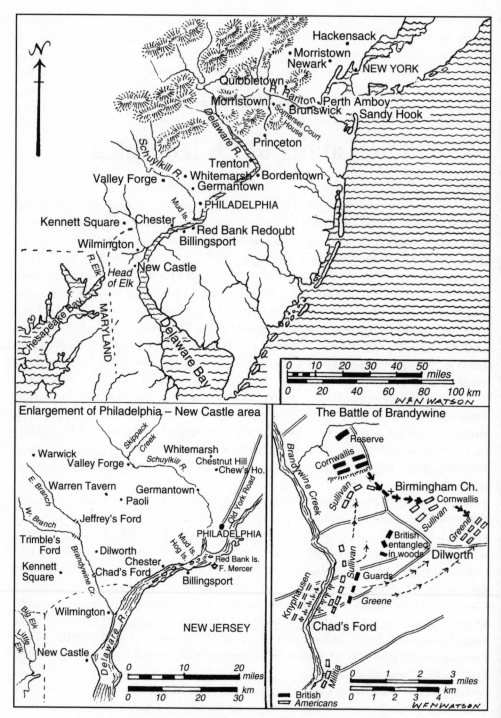

Howe's Invasion of Philadelphia, 1777

other than that a substantial reinforcement of men and equipment was known to be on the way, and that the return of Lord Cornwallis was also expected thereby giving Howe another experienced general under his command and an additional body of troops. The reinforcements arrived on 24 May, and Cornwallis not until 5 June, so that it might be argued that these were factors that justified dilatoriness, but it was to be a further six weeks before the army sailed for the Delaware, and surely a commander of greater energy and resolution would be worried by what might be lurking in the womb of time.

Eventually Howe marched out of Brunswick on the evening of 13 June, and in the course of the night covered a distance of nine miles to Somerset Courthouse, where he remained for several days. The purpose of this manoeuvre was to lure Washington from his strongly entrenched position at Middlebrook, but in this he failed, although he got Washington confused as to his ultimate purpose.[3] Washington's letters to Congress clearly show that he thought Howe was making for the Delaware. So he was further confused when the British marched back to Amboy. This did bring him down from his stronghold with a view to attacking the embarkation and Howe immediately went into action. But the plan to encircle the American left flank, and get between their army and their base, failed when Cornwallis's force met Stirling's and the subsequent fight alerted Washington to his danger. Howe made no further attempt to engage Washington and embarked his troops for Staten Island on 30 June.

There was then to be a three-week delay before the troops sailed to the south. Here again, there were reasons. Howe was anxious to learn of Burgoyne's progress (although at the time he had received no positive orders to co-operate), and he did not want to leave New York until Clinton had arrived from England to take command. Nevertheless, one gets the impression that urgency had been surrendered to the torpor of the times. It was not until 23 July that 260 ships, carrying 13,799 men,[4] left New York harbour and headed for Delaware Bay. There had been a fair degree of secrecy, for the simple reason that the Howe brothers seemed undecided, if not actually divided, as to where the troops were to be landed.

Indeed, among senior naval and military officers there was considerable opposition to going to Philadelphia at all. Foremost among these dissenters was Clinton who, while still at sea, prepared a memorandum for Howe through which he hoped to persuade him to go up the Hudson.[5] In this he was supported by Generals Grey and Erskine, and many naval officers. Cornwallis, however, sided with his chief, who was not to be diverted from his plan. Originally, Howe had favoured going to the Chesapeake, for there were certain strategical advantages and the Delaware was known to be protected by batteries and underwater obstacles. The choice of the Chesapeake had been one of the major objections raised by those opposed to the whole operation, for at that time of year the area was considered to be dangerously unhealthy.

However, on 15 July, information had been received (which proved incorrect) that Washington was moving large numbers of troops towards the Highlands, and it looked as though he would cross the Hudson into New York. As Howe had told Burgoyne that should this happen he would follow, he wrote to Germain the next day saying he would go up the Delaware in order to be better placed to assist Burgoyne.[6] Yet just before sailing on 23 July, he wrote to tell Clinton that he might prefer the Chesapeake.[7] In the event, the fleet did nose its way up the Delaware. However in the afternoon of 30 July, Howe summoned the senior officers to his brother's flagship and reported his decision not to land in that river. The defences were said to be too strong; nevertheless, there would appear to have been no reason why an unopposed landing could not have been made at or below New Castle, which was less than two days' march from Philadelphia. However, on the receipt of the news that Washington was moving on Wilmington, and therefore would not be bothering Burgoyne, it was decided to revert to the original plan and sail for the Chesapeake. These marplot peregrinations of the Howes had upset the local Loyalists and lost an immense amount of time, for they had had to sail 300 miles all the way round a peninsula – but they had at least completely baffled Washington.[8]

The early stages of the Delaware campaign had been marked by those disputations, indecisions, procrastinations and deficiencies of planning which form such a dismal background to many British operations in this war. But now that the fighting would soon begin, Sir William could hope that affairs would fly speedfully upward. The ships had sailed up the bay to Head of Elk, and the troops disembarked on 25 August, only a day's march from the place where they might have landed three weeks earlier. They had been at sea for 42 days, and had suffered a considerable buffeting, but those horses that had survived the journey were in an even worse state than the men. Howe wisely rested until 8 September, while his transports sailed to New Castle ready to victual and support the army. The expected loyalist support proved very disappointing, which prompted Howe to write to Germain on 30 August definitely dismissing any chance of his being able to help Burgoyne,[9] and it also meant that there would be a shortage of garrison troops for conquered land. However, news that Washington had his army entrenched behind Brandywine Creek, some 25 miles to the north, gave Howe the hope of a major victory.

His intelligence had proved absolutely correct. On 9 September he found Washington, with probably around 13,000 effectives, on the north bank of Brandywine Creek. Howe assembled his force at Kennet Square, about five miles from the Americans, whose advance patrols he had driven in, and on the 11th he advanced to contact. Brandywine was a very tidy, well-planned and well-executed battle from which Howe emerged with considerable credit.

Washington had prepared a good defensive position at Chad's Ford commanded by well-sited batteries. To his left, the creek became a virtually impassable torrent running through high banks. His right was held by Sullivan's division. In support, he had Generals Greene and Wayne with regiments well supplied with French arms. A frontal assault was unlikely to succeed, and Howe's plan was for Knyphausen's Hessians to make a feint there while Cornwallis, with the bulk of the army, completed a wide encircling movement to the left to come in on Sullivan's rear. At daybreak on the 11th, the army left Kennett Square; Knyphausen's men made straight for Chad's Ford which they reached about 10am, but Cornwallis had to cover some 12 difficult miles to cross the creek at Trimble's Ford, and its east branch at Jeffrey's Ford.

Shortly after his arrival at the creek, Knyphausen bombarded the American position, and made outward preparations for a major assault, which lured the Americans into sending a strong force across the creek in an unsuccessful attempt to drive Knyphausen off. Washington received contradictory information about Cornwallis's flank march, and final confirmation did not come until about 2pm when he at once moved Sullivan's division to take up a new position at right angles to the main line on some high ground above Birmingham Church. Here Cornwallis's found him at about 4pm and, after a sharp fight, drove him off the position towards Dilworth. But now the densely wooded country took a hand in the fate of the battle; four of Cornwallis's battalions got detached, and were temporarily out of the fight, which seriously weakened his attack. Meanwhile, Washington, skilfully sizing up the situation, sent Greene's division to take up a position on the Chester road covering Sullivan's retreat, from which Cornwallis was not able to shift him until he withdrew under cover of darkness.

As soon as he heard battle joined on the left flank, Knyphausen, whose feint had been successful, now converted it into a thrusting attack across the creek at Chad's Ford. Ably assisted by Cornwallis's 'lost' battalions, who emerged from the wilderness on the right of the American centre at a most opportune moment, the Hessians overran the position and the American retreat became general. However, darkness came before Howe's two columns could join and complete Washington's discomfiture. The British, who had won a complete victory, bivouacked on the field, having casualties amounting to 557 killed and wounded against more than 1,000 Americans, who also lost 400 prisoners.[10]

Armchair critics have again been vociferous in condemning Howe for not pursuing, and thereby losing his best chance in 1777 of destroying Washington's army. Pursuit through dense country with very tired troops was out of the question that night. By the next day, it might well have been too late, for Howe's horses had scarcely recovered from their sea journey and long marching, and anyway the country was unsuitable for cavalry. Howe preferred to spend the days immediately after the battle caring for the wounded and establishing a hospital at Wilmington. It seemed a perfectly sensible thing to do.[11]

For those who like to reflect on the 'ifs' of history there occurred at Brandywine an interesting example of the genre. Major Patrick Ferguson, officially of the 70th Foot, but more notable as a daring partisan leader, of whom we shall hear more later, had George Washington in his sights at a distance of less than 100 yards on two occasions. Washington was reconnoitring with a visiting French officer and rode dangerously close to the British right flank, where Ferguson's men lay concealed. On his return ride, Ferguson hailed him (the Frenchman had gone another way), 'On my calling, he stopped; but, after looking at me proceeded. I again drew his attention, and made signs to him to stop, but he slowly continued on his way.' Ferguson, who was an ace shot, had no idea who the man was, 'but it was not pleasant to shoot at the back of an unoffending individual, who was acquitting himself very coolly of his duty; so I let him alone'. The next day, when Ferguson learnt, from a wounded American officer, the identity of the rider on the bay horse wearing a dark green or blue coat and a large cocked hat, he said 'I am not sorry that I did not know at the time who it was.'[12] Those were the days when, for the most part, chivalry forbade pot shots at commanders-in-chief, but Washington might not have been so fortunate with some.

When Howe did move north towards Philadephia on 16 September, there was a certain amount of marching and countermarching by both sides, with the Americans crossing and recrossing the Schuykill. Washington offered battle at Warren Tavern, but heavy and continuous rain for two days did such damage to American powder that he withdrew across the river to replenish. Howe sent Major General Grey to attack an American rearguard of some 1,500 men and four cannon that were encamped near Paoli, where Grey gave orders for every man to remove the flint from his musket, and the men went in with the bayonet. Surprise was complete. Soon 300 Americans were killed or wounded 'on the spot' and a lot of arms and baggage captured, but Wayne got his cannon safely away.[13] As a result of the action, Grey gained the soubriquet 'No flint'. The road to Philadelphia was now clear and Cornwallis marched in at the head of his troops, to a very mixed reception, on 26 September.

So far the campaign had been a fair success, and Howe's performance nearly foot perfect. It is true that Washington's army, although beaten, was certainly not destroyed, but this might have been accomplished had it not been for the rain, and there was still time. Philadelphia had been captured, and Congress had fled to York, but Howe remained despondent at the lack of loyalist support (hardly surprising with the amount of plundering by his troops), and there had been little or no response to his latest proclamation of 27 August. He knew that Philadelphia sat uneasily in British hands, and that before it could be secure, the Delaware had to be opened for access to the sea. The navy had reported strong defences at Billingsport, and Howe sent three battalions to assist in their reduction. It was also necessary, while the river remained blocked above Chester, for him to detach a strong force to escort supplies from there to Philadelphia.

There now occurred a slight setback born of overconfidence, and not dissimilar to what had happened at Trenton. In the last days of September, Washington, having been substantially reinforced, was moving towards Philadelphia seeking another battle and it seemed that the chance had come when he learnt that Howe had dispersed his force. The latter had sent 8,000 men six miles north of Philadelphia to the village of Germantown with orders not to entrench, yet to be ready for action and to show a bold front, displaying high morale.

Howe, in his dispatch to Germain says, 'at three o'clock on the morning of the 4th the Patrols discovered the Enemy's approach, and upon communication of this Intelligence, the army was immediately ordered under arms.'[14] This is an attempt to disguise the fact that the force had been taken by surprise. The ensuing fight was a tough one lasting for just under three hours, and but for the fortunate occurrence of fog soon after sunrise, and a very stout defence of an outpost known as Chew House, the British, with inferior numbers and taken off balance from non-existent defences, might well have suffered defeat.

Washington's plan was far too ambitious for an inexperienced army even without the handicap of fog, and in the frenzy and turmoil of battle two of his divisions overlapped with unfortunate results. Eventually, his left wing, having suffered casualties from friend and foe, gave way and the whole line was soon in retreat, being 'pursued through a strong country between four and five miles'.[15] Howe had snatched a victory, not entirely deserved, from a battle that developed into a mêlée which greatly favoured British skill with the bayonet. He reported the enemy loss 'between two and three hundred killed, six hundred wounded and four hundred taken. In all 72 officers have been taken since Brandywine.'[16] The British loss was 537 killed and wounded.

Howe was later to excuse himself for not having raised any redoubts for the security of the camp because 'works of that kind are apt to induce an opinion of inferiority, and my wish was, to support by every means the acknowledged superiority of the King's troops'.[17] He further considered it unlikely that Washington would attack so soon after his defeat at Brandywine Creek. In saying this, he seems to have forgotten Warren Tavern, where Washington presented himself eager for battle.

Billingsport had been cleared of enemy before Germantown, but there remained fortifications four miles upstream at Red Bank and Mud Island, which commanded formidable *chevaux-de-frise* that completely blocked the river. Some historians have treated the clearance of these fortifications in a minor key, but in fact they represented some of the toughest work and hardest fighting of the whole Philadelphia campaign, and a passage for ships was not finally cleared until 22 November.

The Delaware bifurcates just above Chester, and only the eastern channel was suitable for warships. At a point where the Schuykill empties into the Delaware, that river is two and a half miles wide and there are five islands between the Pennsylvania and New Jersey banks. Mud Island, in midstream, and Red Bank, with its Fort Mercer on the New Jersey side, were strongly held and supported by a number of armed galleys and floating batteries. Howe's first attempt was to blast the defences on the island with artillery positioned on Province Island; the weather was so appalling, and the mud such an obstacle, that the guns were not in position to open up until 15 October. It soon became apparent that artillery alone was useless. Howe therefore withdrew the troops from Germantown to Philadelphia and made preparations to assault Red Bank. A few days earlier, he had sent an order to Clinton to dispatch 4,000 men to the Delaware urgently.

The assault on Red Bank was made by 2,000 Hessians, supported by five ships, on 22 October. Howe had seriously underestimated the enemy strength, and the Hessians had inadequate cannon and no scaling equipment. Howe had, however, told von Donop, who commanded the force, that he was not to attack if the fort appeared too strong. Von Donop disregarded this order and, in a very gallant but most wasteful assault, lost his own life and suffered 232 casualties to his force,[18] while two ships were also lost. The Hessians were withdrawn, and Howe reverted to the artillery plan, but this time on a larger scale with troops and ships standing by to support. The return of a long spell of ghastly weather caused constant delays, so that the guns were unable to begin their remorseless pounding until 10 November. This time the sustained bombardment brought almost total destruction of the American battery positions on Mud Island and the troops had mopped up by the 16th. Two days later, Cornwallis, with 2,000 men, marched to Chester. Here he was joined by the 4,000 reinforcements from New York. They crossed the Delaware and made for Red Bank, but the Americans had wisely withdrawn upriver in those of their ships not captured. A few days later the navy located and removed the *chevaux-de-frise*. With Philadelphia now secure, Lord Howe could spare more ships for the blockade and he departed in the *Eagle* for Rhode Island.

It had been a messy operation in every sense, but apart from the time wasted in the initial futile bombardment, and in sending von Donop, inadequately equipped, to attack Red Bank, no blame – although not much credit – can be attached to the British command, who as usual had been well supported by the rank and file. Washington's army was now encamped at Whitemarsh, some 12 miles north of Philadelphia. In the hope that a forward movement might tempt him to leave his strong position and give battle, Howe advanced on the evening of 4 December, and took post upon Chestnut Hill on the next morning. Washington detached a corps of 1,000 men against the Light Infantry under Colonel Abercromby, who quickly defeated them. The following day, the van under Cornwallis encountered another enemy force of 1,000 riflemen who were also defeated. But the American position at Whitemarsh was considered too strong for a major attack to be successful, and the force retired to Philadelphia on the 8th.[19]

This ended the season's campaigning. Although Howe considerably outnumbered the Americans, he made no further effort to attack them either at Whitemarsh or in their next encampment at Valley Forge farther west and his troops remained in Philadelphia, most comfortably quartered, until June 1778. As with so many of Howe's actions, there has been considerable controversy about his failure to take the opportunity of destroying Washington's army while at its nadir in the Valley Forge encampment. Howe justified his long period of inaction in a letter to Germain,[20] and there is contemporary evidence that, although the Americans were ill-shod and half-starved, their defences were extremely strong and any attack might have been repulsed with heavy loss.

On the other hand there is also contemporary evidence that an excellent opportunity to finish off Washington's army was missed. He himself, in letters to Congress and Governor Clinton (a senior Congressman who later became Governor of New York) of 23 December 1777 and 16 February 1778 respectively,[21] paints a very clear picture of the desperate state of the American army. Stedman, who was with Howe's army, is most insistent that Howe could have drawn Washington out by a night march or stormed his camp, which was very strong but not impregnable. He asserts that Washington had less than 5,000 men, with frequent desertions of 10 to 50 men at a time, their condition was deplorable, and their cannon frozen up and immovable. An attack, he says, would have boosted Loyalist morale and shaken off the lethargy in which British soldiers had been immersed during the winter. ' . . . had Sir William Howe led on his troops to action, victory was in his power, and conquest in his train.'[22]

Nevertheless, these were times when winter campaigning was not lightly undertaken, especially in difficult circumstances, and Howe's judgement must be respected. Nevertheless, there does not seem any valid reason why an attack should not have been launched in the spring. His excuse then was that he had information that the defences had been considerably strengthened by additional works and he was certain he could move Washington when he opened the campaign.[23] But he never did open it, for as we shall see, he left for England, and his successor was to receive other instructions.

* * *

It is permissible to ask whether the Philadelphia campaign had any justification. It was predicated on the false information that there would be a very considerable loyalist rising, and that Washington's army, in defence of the capital, might be decisively defeated. Even in the unlikely event of these two imponderables being realised (for loyalist support had proved ominously disappointing in the past, and Washington was too wily a general to hazard his whole army unwisely), it is difficult to see how the occupation of Philadelphia by one army and Albany by another could have ended the rebellion. It is, of course, easy to be wise after the event. All one can say is that the campaign achieved extremely little and that it is more likely that the original strategy would have produced better results.

That possibility, and Howe's responsibility for Burgoyne's failure, has been discussed in the previous chapter. The principal criticism levelled against Howe while in New Jersey was not his refusal to attack Washington at Morristown, for it is generally accepted that that would have been a costly folly, but his dilatoriness in the early summer. It is perhaps excusable that he should have wished to wait for the equipment and reinforcements to arrive, for many a commander before and since has disliked going into battle with insufficient troops and equipment, but there were weeks of delay, even after the beginning of June, for which there appears to be no excuse.

There was talk about waiting to hear how the Canada army progressed, and changing the route so as to be on hand to give assistance if needed. The simple fact was that Howe genuinely believed that Burgoyne could get to Albany unaided. Once committed to Philadelphia, he had very little intention of offering Burgoyne any assistance. Howe had a big job to do, and he was determined to do it thoroughly and in his own time. His tardiness and indecisiveness as to the best line of approach probably did not have any serious effect on the outcome of his campaign, but it may have upset potential Loyalist volunteers, who were just as confused over his intentions as Washington was. There was a big lesson to be learnt from the paucity of loyalist support during these, on the whole, very successful operations, but Germain and his ministerial colleagues failed to absorb it.

The Philadelphia campagin was to be Howe's last. The fact that he had achieved so little in Pennsylvania depressed him, for although his leadership in battle could not be faulted, and the army had had some notable successes, Howe knew that it needed more than miltary prowess to suppress a large scale revolt and the disappointing loyalist support had affected him deeply. He was further saddened by the failure attending the considerable energy that he and his brother had expended in pursuit of the *ignis fatuus* of a peace settlement. In October he had tendered his resignation to Germain, which the King at first refused to accept, but on 4 February 1778 Germain wrote an acceptance, and authorised him to hand over his command to Clinton.[24] His brother was to remain in command of the fleet until August, vainly attempting to outman-oeuvre and defeat the Toulon squadron under Count D'Estaing off Rhode Island and Boston, for by then the French, not the Colonists, had become the principal enemy.

The choice of both the Howes to command in America was not, as is sometimes said, a bad one. Both had had experience of the country, both were well regarded by those Americans who knew them, and both were disting-uished in their respective services. However, it cannot be said that Sir William's overall performance during the two and a half years of his command was a great success. There were flashes of excellence, but there were too many imperfections and these latter on occasions allowed Washington and his army to survive. Undoubtedly, both Howes greatly preferred the chance of a peaceful settlement to an attempt at subjugation by conquest, and in this they were encouraged by Lord North's government, whose directives, constantly vacillating between firmness and appeasement, must have been confusing. There is no real evidence that either brother deliberately conducted operations to enable Washington to survive, although this might be a charitable interpretation of some of the general's manoeuvres.

The relationship between Secretary-of-State and Commander-in-Chief was never cordial, but it was not allowed to interfere with the management of the war. Howe was always kept aware of Government intentions and never constrained by interference. His greatest grievance latterly was Germain's refusal to send him the substantial reinforcements (British or foreign) he thought necessary to fulfil his task. He could never believe this reluctance was due to the demands of other enterprises or the difficulties of procurement and he was particularly incensed that so many troops should have been given to Burgoyne.

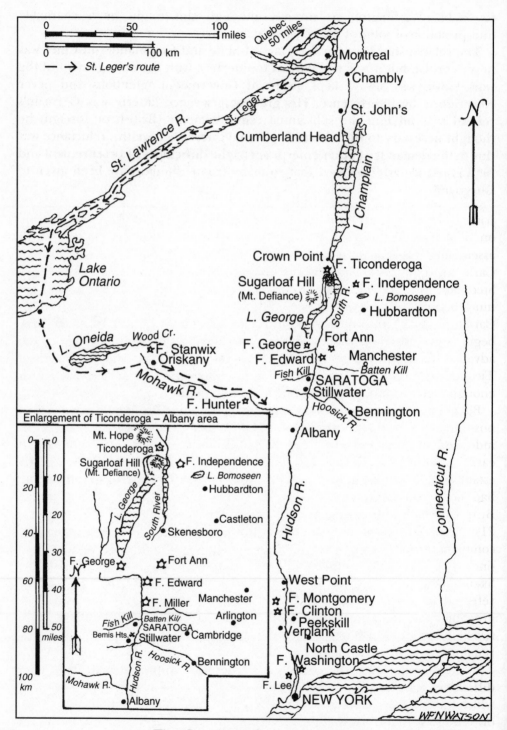

The Saratoga Campaign, 1777

CHAPTER 7

Saratoga: A Good Beginning

JOHN BURGOYNE ARRIVED in Quebec to assume command of the northern army on 6 May 1777. Six days later, he was at Montreal, where his army was assembling. He had been chosen for the Canada command in preference to Carleton, though the latter, in the opinion of many, notably of the historian Stedman,[1] who fought under Howe and Clinton but not under Burgoyne, was much better qualified, with greater local and military knowledge. It is greatly to Carleton's credit that though naturally disappointed, he was from the very beginning most co-operative. The fact that during the planning stage of the advance he was to refuse Burgoyne's request to provide troops to garrison Ticonderoga after its capture was simply because he had not been left with enough men to do so. It had nothing to do with pique.

Burgoyne's objective, in its simplest form, was to get to Albany as quickly as possible to join forces with Howe. Until that point, he would have an independent command. His plan followed the broad outlines of the one he had earlier submitted to Germain in his 'Thoughts' (see Chapter 5) although the establishment and manning of blockhouses on the Ticonderoga-Hudson line had been considerably modified. To enable him to achieve his objective Burgoyne had under his command seven British regiments – the 9th, 20th, 21st, 24th, 47th, 53rd and 62nd. In addition he had eight German regiments, commanded by Major General Baron von Riedesel. The total British force (including 411 artillerymen) was 4,488, the Germans numbering 4,699. At Skenesboro, the army was joined by 148 Canadians, 500 Indians, (usually referred to at that time as 'Savages') and 682 Loyalists. As was usual, the grenadier and light companies were detached from their regiments to form two élite corps to which were added the light companies of those regiments that remained as garrison troops in Canada.[2]

Carleton, who knew the country and therefore appreciated the appalling difficulties for transport, nevertheless advised a strong artillery component, of which the heaviest calibre guns could be left at Ticonderoga once their purpose of reducing that fort had been accomplished. Thereafter, four medium twelve-pounders, two light twenty-four pounders, 18 light six-pounders, six light three-pounders, two eight-inch mortars and four royal howitzers would be transported forward by water and manhandled over appalling tracks. A formidable field train indeed but, according to Burgoyne, the ammunition carried was totally inadequate.[3] And the same could be said of the horses and transport necessary to supply the army, which was a much graver defect than shortage of ball. Given more time, Burgoyne would no doubt have organised his resources better, but to beat the weather he wanted to be at Albany by October.

With the exception of the Indians, who were a constant source of trouble, it was a useful army of some 8,000 combatants to which were attached a large number of non-combatants, including about 200 women. The British element was superbly officered and comprised tough, well-trained and disciplined regular soldiers, while the Germans had spent the greater part of their lives fighting; to them bloodshed had become a profession. Both British and German equipment and uniform, however, were very out of place. The British soldier with his musket, cartouche box, sword belt, knapsack, rations and water bottle humped about 60 pounds over his traditional red coat, waistcoat, close-fitting white breeches and black gaiters, while a Brunswick dragoon stumped around the steaming American forests carrying a broadsword and heavy carbine and wearing tall jackboots, leather breeches and a large hat surmounted by plumes.

Burgoyne, before leaving Canada, frequently expressed seemingly genuine confidence in his ability to achieve his objective, but he made it very clear in letters to Howe from Plymouth before sailing, and again in May from Canada, that he expected the junction of the two armies at Albany.[4] He was disappointed at the inflexibility of his orders, for he wanted to have freedom 'to give all possible jealousy on the side of Connecticut', which meant to make a demonstration or feint into what is now Vermont, if the opportunity presented itself. But in putting forward this suggestion in letters to Germain on 12 May and to Harvey on 19 May,[5] he made it clear that he would not be thus diverted from his principal objective to reach Albany.

On 18 June, at Cumberland Head, Burgoyne issued a proclamation to his assembled troops. Unfortunately it was not the soldier that rose to the occasion but the dramatist, with the result that the troops got a piece of pompous rodomontade that left them gaping and the enemy laughing. Three days later, the Indians were treated to a further display of high-sounding words not one of which they understood. On 25 June the army departed from its base at the River Bouquet in Burgoyne's great fleet of gunboats and bateaux. The army was loosely organised in three divisions or wings. The Advanced Corps, commanded by Brigadier General Simon Fraser, contained the light companies, the 24th Regiment, 60 rangers, and some Canadian woodsmen. The right wing of two brigades was under Major General Phillips, nominally commander of the artillery, but needed for this important tactical command, and whose place in charge of the guns (which were divided among the three wings) was taken by Major Griffith Williams. The Germans formed the left wing and they were organised into two brigades. There was a reserve corps under Colonel von Breymann.

At about the same time, Lieutenant Colonel Barry St Leger set out for the Mohawk Valley. He had 200 British infantry, 100 Germans, 233 loyalists, and 40 artillerymen, together with 900 Indians.[6] This total of 1,473 men was well short of the 2,000 that Burgoyne had specified in his 'Thoughts'. The Indians were led by the renowned Joseph Brant, known to them as Thayendanegea,[7] and a relation by marriage of Sir John Johnson, who was with St Leger, in command of his own regiment, the Royal Greens. Their route was by way of Lake Ontario and Oswego to the Mohawk River, and then east to join Burgoyne.

The latter's advance to Ticonderoga was cautious; he spent three days at Crown Point, and it was not until 1 July that Fraser's Advanced Corps, which had marched from Crown Point, while the remainder went by boat, made contact with the enemy. The American Major General Philip Schuyler, a competent and popular soldier, was in command of the Northern Department, and he had put his ablest lieutenant, Major General St Clair, in charge of Ticonderoga, but the fort had been allowed to fall into a bad state of repair, and St Clair was short of military stores and men. He had only 1,576 Continental troops fit for duty and 300 or 400 militia giving him, with details, a total of no more than 2,200 men.[8] Schuyler had visited St Clair on one or two occasions to offer advice, but neither man showed any sense of urgency, and although 800 reinforcements were sent up on 3 July, this was still inadequate to garrison the whole complex. Even so, this did not excuse St Clair for failing to man the commanding Sugarloaf Hill (Mount Defiance) feature.

Burgoyne had received some information from a militia deserter which exaggerated to 5,000 the strength of the garrison. This reinforced his belief that he would have a siege on his hands, and his plan was to reduce the fort and surround the area to cut off the garrison. Fraser was to seize the fortified mound known as Mount Hope and then swing in against the fort, while Riedesel's Germans marched down the east shore to block the isthmus behind Mount Independence, a feature which was connected to the fort by a bridge of boats. Fraser failed to catch the troops defending Mount Hope, but he drove them out of the old lines and into the fort. By 2 July, he had outflanked the fort from the west and prevented any chance of evacuation along the portage route between Lakes Champlain and George. But the cumbersome Germans had a slower and more difficult route through marsh land, and during 3 and 4 July there was a brisk exchange of cannon and small arms fire between the besieged and the besiegers.

General Phillips had quickly appreciated the importance of Sugarloaf Hill, and Burgoyne's Chief Engineer, William Twiss, had reconnoitred it and found that the guns could be manhandled up the goat paths. Thus, on the morning of 5 July, St Clair awoke to find his position in the fort threatened by British artillery. The combination of this threat and the danger of being surrounded caused him to order a withdrawal to Mount Independence, and when that became unsafe, for the bridge of boats had not been properly destroyed, he marched his men down the east bank of the river towards Hubbardton, while the wounded and stores were dispatched by boat to Skenesboro. He was subsequently court-martialled for evacuating the fort, but he had little choice, and it was entirely right that he was completely exonerated.

The importance of Ticonderoga was perhaps somewhat exaggerated, but Burgoyne could not proceed without its capture, and it had been a well executed and successful operation. Burgoyne, understandably, was delighted, and so was London when news of this seemingly important victory reached the capital at the end of August. The King, in a moment of euphoria, rushed into the Queen's bedroom exclaiming 'I have beat them, beat all the Americans', and Burgoyne was the toast of the town. Howe also may have rejoiced, for Burgoyne's optimistic letters of July may have eased his conscience for having no intention of going up the Hudson to join forces at Albany. And certainly Germain must have been a happier man, for in the knowledge that he had weakly allowed Howe to go his own way, he had written to Knox on 27 July 'if that [Burgoyne's] army is not able to defeat any force the rebels can oppose to it we must give up the contest'.[9]

Unlike Howe, Burgoyne believed in the rapid pursuit of a beaten foe and, apart from two regiments left as a temporary garrison of the fort and Mount Independence, the army gave immediate chase by land and water. The Americans had built a fairly solid boom across the lake just north of the fort which they hoped would defy the British fleet for some while, but by 9am Commodore Lutwidge had broken this formidable obstacle, and Captain Carter's gunboats were quickly after those Americans escaping up the South River. Catching up with their ships at about 3pm, near the Falls of Skenesboro, Carter captured some of the larger galleys, and set fire to many bateaux.[10] Three British regiments had gone with this force, and these were landed once the American ships had been destroyed, but owing to a difficult march they did not reach Skenesboro before the American boating party had set the fort there alight and escaped to Fort Ann, leaving some of their wounded and all of their baggage behind.

There was nearly a minor disaster at this stage of the pursuit when a detachment of 190 men of the 9th Regiment under Lieutenant Colonel Hill became dangerously close to being ensnared by the wiles of an American spy. After a long day's march, made particularly tedious by broken bridges and appalling tracks, the party camped for the night just a mile from Fort Ann. Here they were joined by an American 'deserter', who having fed Hill with false information, disappeared in the night to report to his friends in the fort the strength and condition of the British detachment, which was some eight miles in advance of the main body.

The next morning, the British were attacked by about 400 New York militia under Colonel van Rensselaer, and after a 'smart skirmish' lasting several hours they were in grave difficulty, more especially as some Americans had contrived to work their way round and threaten their rear. Hill withdrew his men to higher ground, and sent back a message to Burgoyne who immediately dispatched Brigadier General Powell at the head of the 20th and 21st Regiments to cover Hill's retreat. But before they could arrive, the situation had become desperate with ammunition almost exhausted. Surrender seemed inevitable. Suddenly, from out of the woods a short way to the north, was heard a series of bloodcurdling Indian war-whoops. The Americans, whose ammunition was also running low, took this to be the leading Indian element of British reinforcements, and made a hasty withdrawal to the fort.

Hill's detachment had actually been saved not by reinforcements but by the initiative of a single British officer, who having been deserted by his Indian scouts, was endeavouring to persuade any American troops in the area that he was not alone! The Americans were shortly to burn Fort Ann and retreat southwards towards Fort Edward; the British force, fortuitously saved from disaster, also withdrew with some 30 prisoners and a militia standard, but Hill had to leave his badly wounded on the ground in the care of Sergeant Lamb and a small guard.

Meanwhile St Clair, marching his Americans at a great pace, skirted round the top of Lake Bomoseen and after 24 gruelling miles reached Hubbardton, where he left a rearguard of the Green Mountain Boys and the 2nd New Hampshires. Fraser's Advanced Corps, followed by the Germans marching at a slower pace than that of the light companies, had left Ticonderoga at 5am on 6 July. Driving his troops relentlessly over appallingly rough tracks and under a blazing sun, he caught up with the American rearguard 24 hours later. These were some of the best American troops and, although taken completely by surprise in the early hours, they fought magnificently and with the utmost courage and stubbornness, foiled every attempt of Fraser's men to drive them from their hill position. Only the timely arrival of von Riedesel's advanced troops overwhelmed and dislodged these sturdy soldiers. The action had lasted 45 minutes and cost the British and Germans 174 casualties, while the Americans lost 12 officers and 283 men, of whom 228 were prisoners.

St Clair's force was now in disarray, two militia regiments ordered to turn back to reinforce the Hubbardton troops refusing to go in to battle. All he could do was to collect what troops he could and, making a wide circuit to the east, join Schuyler at Fort Edward. His arrival there on 13 July brought Schuyler's force up to about 3,000 but they were mostly dispirited men, short of provisions and with little protection against the heavy rains that had recently developed. Schuyler appealed to Washington for reinforcements, but the Commander-in-Chief was still puzzled as to where Howe was going, and had to remain strong in New Jersey. Burgoyne, on the other hand, was ebullient; everything had gone splendidly, and his friend – but dangerous counsellor – Colonel Skene not only entertained him sumptuously in his house at Skenesboro, but assured him that the considerable number of Loyalists who now rallied to him was but a fraction of those who would join farther south.

It would seem that Burgoyne had Schuyler's demoralised troops, now concentrated 20 miles away at Fort Edward, at his mercy. But it was important that they should not be given time to rally, re-equip and be reinforced, and Burgoyne was losing time through lack of transport and horses for the heavy guns, which he was certain would be necessary. He was now faced with a difficult choice of routes, each of which had its own advantages and disadvantages. That he chose correctly seems fairly certain, but there were – and still are – those with serious doubts. As it might have affected the outcome of the campaign, it is worth looking into the alternatives.

He could take the direct route overland. Skene recommended this (although there were some who said he wanted the track improved for his own benefit) and Twiss, who had carried out a reconnaissance, reported that the many problems were not insuperable. Nor were they, but it was at best a narrow, low-lying forest track, often swampy and always very difficult for transport; moreover there were many ravines and the bridges were likely to have been destroyed by the retreating Americans.

The alternative was to go back by the South River and Lake Champlain to Ticonderoga, carry his boats across to Lake George and sail for Fort George at the end of the lake. The short portage between the two lakes was over a reasonable track, and the distance between Forts George and Edward was only 10 miles over a good track. This was undoubtedly a longer route, but there were those who thought it could be quicker, and indeed it was the route to Albany that Burgoyne had originally suggested in his 'Thoughts'.[11] However, in his Narrative he gave many reasons for discarding it.[12] Among these were the loss of morale, especially among the Loyalists, in going backwards; the need to reduce Fort George; the loss of a possible opportunity to make a feint into New England – a frustrated ambition Burgoyne regretted in a letter to Germain of 11 July[13] – and when the Fort George garrison eventually withdrew they would proabably destroy the road.

The decision was therefore made to take the direct route, and to send the heavy baggage back by boat. This was excessive, for the officers – especially the Germans – did not go to war lightly accoutred. It is impossible to gauge how much time – if any – might have been lost by going back, for there are so many imponderables connected with a course not adopted; Burgoyne, himself, made no forecast, but his deputy-quartermaster-general, Captain John Money, giving evidence at the Inquiry said 'it therefore would have delayed the army a fortnight longer than they were delayed to have returned from Skenesboro to Ticonderoga, and gone across Lake George.'[14] This seems an excessive estimate, considering that it took nearly three weeks to accomplish the overland march at the restricted pace of only one mile a day. But Money was more familiar with local conditions than it is possible to be at a distance of over 200 years.

In his Saratoga Narrative, Burgoyne gave the need to reduce Fort George as a reason for taking the longer route. He also made the point that, fearing to be cut off by his advance, the Patriot garrison had withdrawn without a fight. This may have been a bonus, but if Lieutenant Digby, who marched with the 53rd Regiment, is correct in his journal that Fort George was known to contain a large quantity of wagons, horses and stores,[15] it must have been a mistake by Burgoyne not to have sent a flying column under Fraser to seize them – but again we know too little of the conditions at the time to be certain.

Burgoyne's troops started their hard slog to Fort Edward on 11 July. Schuyler had done what he could to make their march intolerable: the bridges had been destroyed; in the denser parts of the forest the narrow tracks had been blocked by the felling of trees; streams had been dammed to cause morasses; crops and farms had been burnt, and livestock totally removed. Burgoyne had hoped to use his Indians to spread terror in advance of the army, and if possible prevent the worst of the American scorched earth policy, but apart from a few valuable scouts they proved unreliable and uncontrollable. Friend and foe might fall equally a prey to their hatchets, and there was one horrendous incident when a girl called Jane McCrae, engaged to marry a Loyalist who marched with Burgoyne, was waylaid and scalped by a party of Burgoyne's Savages as she awaited her fiancé near Fort Edward. No one was more upset (except the fiancé) than Burgoyne, but it did him much harm, for it was a propaganda gift that the Americans exploited to the full, and it was at least partly responsible for the almost total absence of Skene's promised Loyalists.

The main body of the army arrived at Fort Edward on 30 July, which post Schuyler had already abandoned. Morale was still high after the punishing march, but the troops were extremely weary, and glad of the chance to rest while supplies caught up. Logistics were a constant problem to Burgoyne, and numbers were rapidly becoming one. He could not advance beyond his supplies, which were at the mercy of unreliable Canadian contractors, and he was well aware that time was not on his side. Schuyler's defeated troops may have been tired and dejected after their long withdrawal, but he could expect substantial reinforcements, while Burgoyne's army was being constantly diminished. Not only were the Indians beginning to desert but the further his army marched, the more men he had to find to garrison the empty forts along his route. Furthermore, sickness was taking its toll of his ranks. Intercommunication was a tedious and unreliable business, for all letters and dispatches had to go by boat. All this time Burgoyne had no knowledge of what Howe was doing, nor had he news of St Leger beyond the fact that he had reached Fort Stanwix. In such circumstances, the commander of an army is a very lonely man.

Schuyler, too, had pressing problems. On arrival at Fort Edward, he had only some 1,500 men and no idea as to the whereabouts of St Clair's force. He had made Burgoyne's march as difficult as possible and now sent a detachment to the commander of Fort Ann to encourage that officer to offer a stern resistance to the British. However this the latter failed to do and the fort was abandoned without opposition. On 13 July, St Clair's force of weary men turned up at Fort Edward, bringing Schuyler's numbers to around 4,600. Nevertheless, he still did not feel strong enough to confront Burgoyne there and withdrew his men to a defensive position north of Stillwater.

Having been apprised by Schuyler of his difficulties, Washington was certainly not indifferent to the many trials and tribulations the commander of the Northern Department had to bear. He knew that Schuyler was being bitterly, and most unfairly, attacked in Congress for having lost Ticonderoga, even to there being talk of treachery; he therefore sent Generals Lincoln and Arnold to assist him, with a promise of reinforcements from New England. But, numbers count for little if morale is low. The American soldiers had had just about enough withdrawing when Schuyler took them another 12 miles down the Hudson from Stillwater. At the end of the month, Congress decided to replace Schuyler with General Gates: certainly Schuyler was tired, and a change of commander can often lift morale, but it was a bad decision, for Schuyler was an incomparably better general than Gates. One of his last actions in command had been to answer an appeal from Colonel Gansevoort, who was beleagured by St Leger at Fort Stanwix. Schuyler sent Arnold with 900 soldiers to his assistance when he knew full well that Burgoyne, still with superior numbers, was poised to cross the Hudson and attack him. It was a bold and generous gesture, displeasing some of his officers, who felt he should have put his own fortunes foremost – as many generals would have done.

The crisis point of Burgoyne's campaign was now at hand. In the absence of any definite information about Howe, he had a number of options open to him. He had written to Germain the day he arrived at Fort Edward saying 'I have spared no pains to open a correspondence with Sir William Howe, but of the messengers sent out at different times by different routes, not one is returned to me, and I am in total ignorance of the situation or intention of that general.'[16] What Burgoyne was not in total ignorance of was the deplorable state of his supplies of every kind, including horses which von Riedesel was particularly anxious to procure for his dragoons, who were floundering about uselessly in their jackboots. Burgoyne therefore lent a willing ear to his friend Philip Skene, who told him that the area round Manchester and Arlington in Vermont was rich farm land, and so far untouched by war, with good stocks of corn, cattle and even horses. Furthermore, there were large numbers of men there willing to take up arms for the King.

The choices open to Burgoyne were: to remain at Fort Edward until the situation became clearer; to advance rapidly while Schuyler's men were still off balance, and hope to drive them back to Albany before they had time to consolidate their new position; to strike east into Vermont to 'give a jealousy', or perhaps go east and south, bypassing Schuyler's army and making for New York; to go back to Ticonderoga while there was still time and the route still open; or to remain in the vicinity of Fort Edward while a detachment struck east to raid the rich pastures of Vermont. Burgoyne, urged by Skene, chose the last of these options. Although there are those ready (especially with hindsight) to condemn his decision, and eager to advance one or other of the possibilities as being better, it is difficult to say that he was wrong. Had he known that Howe would abandon him he might have chosen differently, but it is doubtful. At all times, he was at pains to emphasise that his orders were inflexible, and he had to get to Albany. Moreover, he had to keep faith with St Leger, who was fighting his way up the Mohawk Valley. A bold general might have taken a chance, and led his men forward, but Burgoyne was too cautious to advance into enemy territory short of vital supplies and, in the circumstances, he was surely right.

Riedesel had instigated the idea of a raid soon after Ticonderoga. He wanted it to be a fairly small tip-and-run affair, but Burgoyne would not agree, and insisted on an operation in some strength. He did, however, choose the Germans to undertake it, possibly because they were nearest to the objective, or because it was Riedesel's suggestion. But if so, neither of these reasons could justify sending heavily encumbered dismounted dragoons crashing through the undergrowth when Fraser had men available noted for their mobility. According to Stedman, Fraser made a strong remonstrance to Burgoyne on this point, but the general remained adamant.[17]

The force that Burgoyne and Riedesel decided upon, in addition to the 200 Brunswick dragoons, was 50 British marksmen (rangers) under Captain Fraser, about 150 Loyalists and Canadian Volunteers, some Hesse-Hanau gunners with two three-pounders and 100 Indians making a total of about 500 men.[18] Stedman says that a Loyalist with local knowledge, who was to accompany the expedition, urged Burgoyne to increase the numbers very considerably,[19] but again he remained adamant. To command the force Riedesel selected Colonel Baum, an exceptionally brave and reliable officer, although without a word of English and, of course, no local knowledge. To a certain extent these deficiencies were made good by the inclusion of Skene and the local Loyalist, but as neither spoke German an interpreter was needed. Events were to prove that the Loyalist was right, for indeed the numbers were insufficient, necessitating the sending of reinforcements. Burgoyne then compounded his original error of composition by dispatching more Germans to plod through the jungle at a snail's pace.

* * *

The choice of Burgoyne to command the northern army in preference to Carleton or Clinton has often been criticised. All such appointments are political, and this one was particularly so, for it is virtually certain that Carleton would have got the command had Germain not been the Secretary of State responsible. Nevertheless, Burgoyne was not a bad choice, and he did very little wrong. Weighty decisions had to be taken, and of course some mistakes were made. These, with hindsight, can be seen to have affected his chances adversely.

There was little Burgoyne could have done towards the initial composition of his force, or towards solving the problem of the ridiculously accoutred and horseless German dragoons, which was eventually to influence his decision to mount the disastrous raid into Vermont. The large artillery component that accompanied the army as far as Ticonderoga was partly necessary for the subsequent defence of the fort, and thereafter the expectation of battle made it necessary to have a strong artillery arm, but in the interests of speed logistics were to suffer. Ticonderoga was well planned and executed, and the immediate pursuit efficiently conducted, although the dispatch of 190 men into unknown enemy held country no less than eight miles ahead of the main army seemed to be asking for trouble – and very nearly got it.

The question of the choice of route for the army to take between Skenesboro and Fort Edward has been covered. The probability is that Burgoyne made the right decision, even though Lieutenant Digby tells us it was not popular throughout the army. There was talk of 'going back' and 'flying columns', but Burgoyne was never one for going back – even temporarily as at Skenesboro – and was inclined to base this resolve on the inflexibility of his orders, which was nonsense and led to what was probably his worst mistake of the campaign. As for flying columns, these do not usually fulfil their function when faced with blown-up bridges, blocked jungle tracks and near impassable swamps. A quick, decisive blow against Fort Edward was never a viable option, but it is possible that failure to take Fort George with its valuable booty was a missed opportunity.

The only serious error that Burgoyne made at this stage of the campaign concerned the composition of the force for the raid into Vermont. It has been shown that he had various options, and that he was probably right in choosing to call a halt while he sent troops for what he was assured would be the easy procurement of badly needed supplies and horses. But why burden poor Baum with a force that included dismounted, jackbooted Brunswick dragoons, an odd assortment of Tories and Canadian woodsmen, a number of women camp-followers and a party of musicians on what was designated a 'secret expedition'? It is possible that von Riedesel, who was the principal advocate of the raid, had been given a free hand. If so, Burgoyne should have overruled his choice of personnel. A small, fairly mobile force of, say, British and German light companies, together with Captain Fraser's marksmen, would probably have had a much better chance of success on the information available than the cumbersome, disparate collection of men and women selected to go tramping through the thickets. And when that information turned out to be hopelessly wrong, they could have withdrawn without loss.

CHAPTER 8

Saratoga: A Distastrous Ending

COLONEL BAUM'S ORIGINAL orders from Riedesel were to leave Fort Miller on 12 August for a raid on the Manchester-Arlington-Connecticut river valley areas. However, shortly before he was due to leave, specific information was received concerning a large supply of horses, food and stores at Bennington, West Vermont, guarded by only 400 militia. Orders were therefore sent to Baum that he was to march direct to that place, necessitating a change of direction from east to almost due south. Riedesel was unhappy at this, for it exposed the flank of this small force to Schuyler whose troops were concentrated just across the river, but Burgoyne discounted it and believed that there would be little or no opposition. Nor was there from Schuyler, but Burgoyne's general optimism was to be sadly shattered, largely due to that John Stark whom he had met at Bunker Hill two years earlier. After Trenton, Stark had returned to New Hampshire. There he had resigned his commission on learning that Congress had promoted juniors over his head, but Burgoyne's invasion quickly brought him back into the fold with a commission to raise a brigade of militia.

Baum marched on the road to Cambridge (not to be confused with Cambridge, Massachusetts), which was little better than a rough track through the forest, and the first 16 miles took his force 12 hours; slow going even for dismounted dragoons. Somewhere in the vicinity of Cambridge he clashed with a small American force and took some prisoners, from whom he learnt that Bennington was held not by 400 militia, but by more than 1,500 well-armed troops. These were mostly Stark's men, for that controversial and contrary officer, having gained his way with Congress that his New Hampshire brigade must be independent, had recently enlisted 1,492 all ranks and was marching them to Bennington, where the military depot would take care of their needs. He was shortly to be reinforced and, with a total of 2,000 men, had a three-to-one superiority over Baum.

Early on 14 August, Baum had another brush with a small American detachment where the Owl Kill joins the Hoosick river. These men offered little resistance but, in their withdrawal, destroyed a bridge which further impeded Baum's cautious progress along the track to Bennington, Later that same day, the Germans sighted Stark's force. Baum was not prepared to offer battle and Stark withdrew his men. Baum had by now sent two messages to Burgoyne stating, among other things, that his Indians had become uncontrollable and that he urgently required reinforcements. Burgoyne received the call for help early on 15 August, by which time the army was at Fort Miller. He immediately sent Colonel Breymann with the Brunswick grenadiers, light infantry, chasseurs and two six-pounder guns, which were carried with their ammunition on carts.[1]

The 15th of August was very wet and Stark was unable to make use of his commanding firepower. Baum used the respite to make what defensive preparations he could. The Bennington road ran east from the Hoosick river and parallel to the Waloomsac river, until the latter curved westwards at the foot of a prominent eminence that rose some 300 feet above the swampy river valley. The hill afforded a very fair defensive position. Had Baum decided to hold it in depth behind what abatis could be constructed in time, he might well have kept Stark at bay until the arrival of the reinforcements he had been told were on the way. But he decided to disperse his force in small groups some on various parts of the hill, with some on the low ground on both sides of the road, thereby subjecting them to being overrun in penny packets.

John Stark was a natural leader with the courage to take risks and the magnetism to fire his troops. He devised a bold plan that required very careful co-operation, for it involved extensive detours over difficult country. He split his force into two principal components in order to carry out a pincer movement, with a third force to act as a frontal diversion. That would be developed into a full attack when the two wide encircling movements were in a position to open the battle. It certainly lacked the merit of simplicity, and it was perhaps fortunate for Stark that Baum was unable to exploit his wide dispersion.

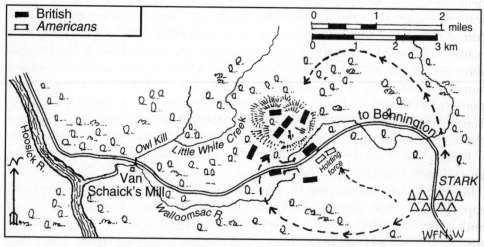

The Battle of Bennington, 16 August 1777

It was not until 11am on 16 August that Stark considered the weather had cleared sufficiently for him to despatch his two flanking forces on their long trail to the north and south of Baum's position. It was after 3pm before the northern wing was in position to open fire, which was a preconceived signal for the whole attack. The outposts were overrun very quickly, but the fighting on the hill was some of the fiercest of the war. Stark's men stormed the steep, rock strewn slope, and his marksmen had soon picked off the German gun team. Then, for two hours, it was hand-to-hand grapple with neither side giving ground. When at last most of Baum's men had found the pace too hot, his dragoons stood by him. Unsheathing their sabres, they carved their way through the less heavily armed Americans. But the numbers against them were too great. When their gallant leader fell mortally wounded, they, like the others, melted away into the woods.

It was a spectacular victory for Stark, but it might have gone very wrong had Breymann marched his 640-strong relieving force with less caution and arrived on the battlefield in time. He had 25 miles to cover, the road was deep in mud, it was raining hard, carts were overturned, a river had to be forded and, worst of all, his guide lost the way. These were certainly valid excuses, nevertheless, in view of the urgency of his task, he might have improved on the first day's march of half a mile an hour, if by no other means than by dispensing with the display of Teutonic thoroughness in halting the column at frequent intervals to dress ranks, He eventually arrived near the battlefield, having had to contest a few minor skirmishes, a short while after the last of Baum's dragoons had disappeared into the wilderness.

Stark's men were scattered, enjoying the fruits of victory, but Seth Warner's Green Mountain Boys, who had not arrived in time for the battle, now appeared upon the scene. This was as well for Stark, because without them he would have been in difficulty, although Breymann was to say later that had he heard of Baum's defeat 'I certainly would not have engaged the enemy'.[2] As it was, the fight was fiercely contested, for Breymann was a stout-hearted commander, but with no experience of forest fighting. His cannon did considerable execution but, with his flank turned, and his ammunition running short on account of bad fire discipline, he broke off the engagement at about sunset. His withdrawal soon took the shape of a near rout, with Breymann himself badly wounded. However, darkness came just in time. In his report, Stark stated, 'But had daylight lasted one hour longer, we should have taken the whole body of them.'[3]

Burgoyne's losses at Bennington were extremely serious. It is impossible to give accurate figures, for every account – contemporary and later – differs. The total was almost certainly over 600, of which the Germans lost 527 killed or captured including 28 officers. The Americans, who lost only about 40 killed and as many wounded, also captured four brass field pieces, 250 dragoon swords and a fair quantity of muskets.[4] Burgoyne's army now consisted of 6,074 men fit for duty of whom 4,646 were regular soldiers.[5] On 20 August, he wrote an interesting private letter to Germain, which began with a justification for ever having embarked on the operation, and a plea that 'your Lordship will, in your goodness, be my advocate to the King, and to the world, in vindication of the plan'. This is followed by a masterly meiosis, 'The consequences of the affair, my Lord, have little effect upon the strength and spirits of the army', but the letter continues in varying degrees of despondency. Messengers to and from Howe, he says, were clearly not getting through – he knew of two that had been hanged – 'only one letter has come to hand, informing me that his intention is for Pensylvania'. And so Burgoyne now knew that he was unlikely to find help in Albany. Nevertheless, after glooming on the deplorable loyalist situation ('The great bulk of the country is undoubtedly with the Congress') and lamenting the fact that as no threat had been made to the Highlands, Gates had been able to build up an army greatly superior to his, he ends on a note of cautious optimism. 'I yet do not despond – should I succeed in forcing my way to Albany, and find that country in a state to subsist my army, I shall think no more of retreat, but at worst fortify there and await Sir W. Howe's operations.'[6]

Burgoyne had scarcely recovered from Bennington when news came of St Leger's failure in the Mohawk Valley. It will be remembered that he had set off from Montreal in June to form the right hook of Burgoyne's invasion from the north. He had with him 1,800 men, more than half of whom were Indians, whose loyalty or defection could be the catalyst of victory or defeat. His first objective was Fort Stanwix which guarded the portage between Wood Creek and the Mohawk River, and which was defended by Colonel Gansevoort with men from New York and Massachusetts regiments. A siege became necessary and there was some delay while St Leger awaited the arrival of his four cannon and mortars. Meanwhile, on receiving news of St Leger, Colonel Herkimer, of the Patriot county militia, marched 800 men to relieve the fort and requested Gansevoort to make a diversionary attack. St Leger, learning from an Indian scout of Herkimer's approach, laid a very successful ambush at Oriskany, some six miles from the fort. And on 6 August, Herkimer's men were only saved from total elimination by the arrival of the diversionary force. In the battle that developed, St Leger's Indians found close-quarter fighting not to their liking and most of them took to the woods.

At the end of the day, Herkimer, although mortally wounded, and with only a few men left, still held the field. Nevertheless he had failed to relieve the fort and so St Leger claimed the victory. But his Indians, who had been further demoralised by finding that their camp had been looted in their absence, were agitating to be away. Before St Leger had time to starve the garrison out, Benedict Arnold had marched to its relief with the 900 men under Brigadier Ebenezer Learned, generously spared by Schuyler. By a clever ruse as to his numbers and the timing of his aproach, Arnold spread such alarm among the Indians that they refused to fight any longer. This so depleted St Leger's force that, on 23 August, he had no alternative but to retrace his steps to Canada.

The recent setbacks at Bennington and Fort Stanwix, together with the near certain knowledge that Howe would not be coming to his assistance, presented a different situation to the one that Burgoyne had faced at Fort Edward at the beginning of August. His supply line was very erratic, his Indians were deserting, the Loyalists were proving no more than a pleasant chimera, and there was no news from Clinton, who anyway had insufficient troops to be of much assistance. There seemed every reason for him to go back to Ticonderoga, or at least to retire to Fort Edward and await events, for by now the odds were stacked heavily against him. Indeed, that is what Germain expected him to do, as he made clear in a letter to Under-Secretary Knox when news of Bennington was received in London,[7] although, by that time, Germain was already attempting to shift any blame there might be on to military shoulders.

If Burgoyne gave any thought to going back, it must have been transitory, for he would have considered there were many reasons why such a course was undesirable. Foremost among them was what he had written to Germain in the letter of 20 August, partly quoted, 'my orders being positive to "force a junction with Sir William Howe."'[8] This was an obsession with Burgoyne, and he seems to have forgotten or misunderstood that sentence in his instructions received from Germain via Carleton, 'until they [Burgoyne and St Leger] shall have received orders from Sir William Howe, it is His Majesty's Pleasure that they Act as Exigencies may require . . .'[9] No orders had been received from Howe, and in any event a general officer in Burgoyne's position, more or less incommunicado and fully aware of the dangers, could be expected to use his own initiative and take what action he considered best to safeguard his army. But probably he remembered his own general order issued in June at Crown Point, which ended, 'This Army must not Retreat', and with high regard for honour he would have been loath to renegue on that rash edict. Moreover, he was a gambler with a great desire for military glory, such as would be his alone if he attempted the absurd and achieved the impossible.

And so the die was cast, and with the opportunity for honourable withdrawal allowed to pass, the army would go forward to Albany. It was a decision that has been frequently, and rightly, criticised. On 13 September, the British troops crossed the Hudson by a bridge of boats placed just above Saratoga, and the Germans crossed the next day. The army, now augmented by some reinforcements that had joined just before the crossing, marched in three columns down the west bank of the river. The mood of these men was valiant, but many perils lay ahead.

When Gates, a cautious and unimaginative officer, succeeded Schuyler in command of the American army, he took his time in deciding where best to halt Burgoyne. He eventually settled on a place a mile or two upstream from Stillwater, where the Albany road passes through a defile between the Hudson and the steeply rising wooded ground immediately to the west. Here Gates found a naturally strong defensive position, which, through the prowess of a Polish engineer called Kosciusko, was made even stronger. The position,

known as Bemis Heights, after Jotham Bemis who owned a nearby tavern, was
in the region of one and a quarter miles in length and three-quarters of a mile
in depth, and there was a deep ravine on its immediate front. Gates had under
command 28 regiments of foot, 200 light cavalry and 22 cannon.[10] He also had
Arnold who, unlike Gates, possessed a natural and instinctive affinity for the
front line. His continental soldiers and Colonel Daniel Morgan's 500 marksmen
were first class, but the New England militiamen were very raw. With a total of
some 9,000 men, Gates was numerically superior to Burgoyne, was soundly
barricaded and time was on his side.

Gates had good information on the progress of Burgoyne's army, for his
deputy-adjutant-general, Colonel Wilkinson, had carried out a close reconnaissance through the thick country. Furthermore, some Germans had
defected to him. Burgoyne, on the other hand, was advancing almost blind. By
now his best Indians had gone and there was a long stretch of well wooded
country between him and Bemis Heights. He was certain, however, that Gates
would have blocked the Albany track at the defile, and had received what
appeared to be a reliable report that, on the left of the enemy position, there
was an unoccupied hill. He decided, therefore, not to continue down the
riverside track and engage Gates frontally, but to feel for the left of the position
and try to turn it, and roll the line off the bluff towards the river. His plan was
sensibly simple: Fraser's troops, together with Breymann's reserve corps, would
sweep round to locate the American left, General Hamilton, with four British
regiments and a brigade of artillery would attack in the centre, while Phillips
and Riedesel with the left wing would remain on the river road.

The army left the Sword House in thick mist and drizzling rain at 8am on 19
September, but the mist had cleared an hour later and the sun shone brilliantly
on the red coats and gleaming bayonets of Burgoyne's heavily accoutred soldiers,
as they weaved their way through the unaccustomed conditions of bush and
forest. Burgoyne led the centre column along a wagon track that crossed the Great
Ravine and then continued westwards to a point just north of the unoccupied
Freeman's Farm. Freeman was a Loyalist, who, with two or three other subsistence
farmers, had hewn a clearing out of the forest, and was to give his name to the
forthcoming battle. Here Burgoyne halted to allow Fraser's column to reach its
starting position. Shortly after 1pm, a pre-arranged minute-gun signal indicated
the time for the widely spread columns to advance to contact.

Gates, characteristically, was not in favour of offensive action; he had
planned to remain behind his defences and let the British attack him. But
under pressure from Arnold, commanding the left wing, he relented so far as
to allow Morgan's riflemen, supported by Dearborn's light infantry, to go
forward. These men engaged the forward troops of Burgoyne's centre column
and virtually wiped them out. Battle thus joined was soon to become a savage
pounding match that lasted until nightfall. Morgan's troops were reinforced by
two regiments from General Poor's brigade, but attempts to work round the
British right were unsuccessful. The brunt of the fighting was borne by the

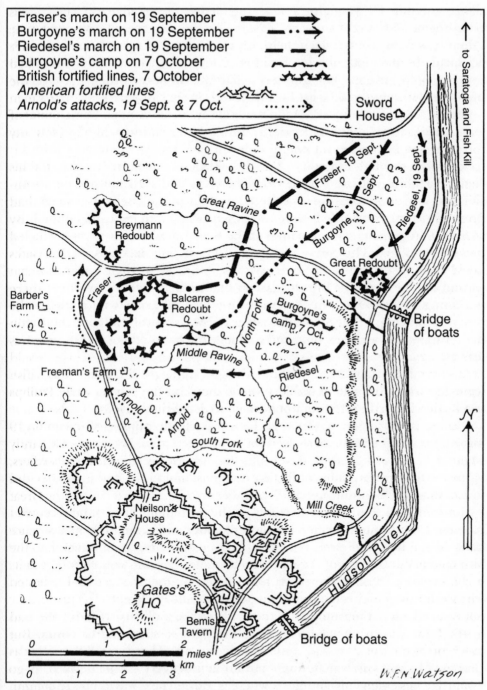

Fraser's march on 19 September
Burgoyne's march on 19 September
Riedesel's march on 19 September
Burgoyne's camp on 7 October
British fortified lines, 7 October
American fortified lines
Arnold's attacks, 19 Sept. & 7 Oct.

Sword House

to Saratoga and Fish Kill

Fraser, 19 Sept.
Burgoyne, 19 Sept.
Riedesel, 19 Sept.

Great Ravine

Breymann Redoubt

Great Redoubt

Barber's Farm

Fraser

Balcarres Redoubt

North Fork

Burgoyne's camp, 7 Oct.

Bridge of boats

Middle Ravine

Freeman's Farm

Arnold

Arnold

Riedesel

South Fork

Neilson's House

Mill Creek

Hudson River

Gates's HQ

Bemis Tavern

Bridge of boats

0 1 2 miles
0 1 2 3 km

W.F.N Watson

The Battles of Freeman's Farm, 19 September, and Bemis Heights, 7 October, 1777

centre column. Of Hamilton's four regiments, the 62nd took tremendous punishment and fought superbly, earning Burgoyne's special commendation. Fraser gave the centre what help he could, but Morgan kept him fully occupied. In the close-quarter fighting, the British gained little from their superiority in cannon. The gunners – like the officers in their eye-catching plumage – were a special target for the American sharpshooters.

On the American left wing, Arnold was always prominent and his troops did well. Gates, fearing his impetuosity, refused to reinforce him, which was fortunate for Burgoyne, for one more push and Arnold would very probably have broken through to victory. As it was, Riedesel's men arrived just in time; General Phillips had ridden up from the river and seen how desperate the situation was, and he personally ordered up four of Riedesel's guns to take the pressure off the 62nd. Riedesel himself then swung into the battle at the head of his own regiment, two companies of the Brunswick von Rhetz regiment and two six-pounders. This heavy punch, and fast-falling darkness, undoubtedly saved Burgoyne's centre. The Americans withdrew to their entrenchments leaving the battle-weary British masters of the field, but with 600 casualties (including 35 officers). The Americans lost 283 all ranks with a further 33 reported missing.[11] Burgoyne himself was always in the thick of the battle, delivering orders with precision and coolness, but he could not do very much to influence events.

At every important stage of the Saratoga campaign, Burgoyne was called upon to make a difficult decision, often without much information. Unfortunately, he was a man who found it difficult to decide and favoured a discussion and consensus of opinion, although he was perfectly capable of overriding a majority vote. The immediate decision to be taken after Freeman's Farm was whether to attack the next day while the Americans might have been off balance. Officers gave him conflicting opinions, and in the end it was decided to wait until the following day (21 September).

However, either early that morning, or the preceding evening, he received a message from Clinton, written on 11 September and brought to him at great peril, to say he could probably spare 2,000 men to make an attack on Fort Montgomery at the end of the month. The messenger was sent back that same night, urging Clinton to any effort that would ease the pressure on Burgoyne, who would await events in his present position until 12 October.[12] The decision not to attack on the morning after the battle was probably correct, for the bulk of the army was in no state to resume the fight so soon, but to postpone it indefinitely on the 21st was undoubtedly wrong. There was still a slender chance of fighting his way through, but for almost every day he delayed, Gates would be increasing his strength, whereas Burgoyne's was diminishing and food was getting short. There was nothing to be gained by waiting on Clinton's limited attack over 100 miles away.

In his letter to Clinton, Burgoyne asked for his 'most explicit orders, either to attack the enemy or to retreat across the lakes while they were still clear of ice'.[13]

Clinton very properly refused to give him any orders, although, without saying so, he favoured retreat.[14] Riedesel, and many officers, were in favour of this course, but Burgoyne again turned it down, although at the time he did not know that it was no longer a viable option. The backdoor had been firmly closed by a force under General Lincoln that included the troops of John Stark and Colonel John Brown, which a few days after Freeman's Farm had re-taken Fort George and captured many of Burgoyne's supply ships on Lake Champlain.

Having failed to seize the initiative, and decided against going back, it only remained to strengthen his line, which now ran from the Hudson to the high ground north-west of Freeman's Farm, and to await events. A series of redoubts, protected by palisades and earthworks formed the principal pivot points of the line, the strongest of which was named Balcarres Redoubt after the commander of the light and grenadier troops. Almost equally strong was the Great Redoubt, which overlooked the Hudson and the river road, and offered protection to the hospital, artillery park and supply depot. Echeloned back from these was the Breymann Redoubt, and as much use as possible was made of natural obstacles along the whole line.

On 3 October Clinton fulfilled his promise to attack the forts. He embarked 3,000 troops in 60 vessels and, escorted by a naval squadron, sent them up river. They captured the two forts, Montgomery and Clinton (named after General Clinton's father, a former Governor of New York), that barred the way up the Hudson just above Peekskill, inflicting heavy casualties on the Americans, and burnt the flotilla which was unable to escape owing to contrary winds. Clinton had made it clear to Burgoyne in one of the many messages that winged their dangerous way between the two men that he had not sufficient troops to fight his way to Albany. However, after he had overcome the forts, this was not the case, for he must have known there was then no enemy force of any consequence between him and Gates. His comparatively easy success did embolden him to go on up the river, but when within 40 miles of Albany the river pilots for some reason refused to take the vessels farther. A feeble ending to early promise.

On 3 October, Burgoyne cut the troops' rations,[15] for it had become clear that supplies would not last beyond the month, and two days later he held a council of war. At this council he gave the officers a resumé of Clinton's situation, not having yet heard of his success, and of Howe's activities; then told them that it now seemed clear that from the first, it was intended that his army could be sacrificed in the interest of other operations.[16] The council then considered options. Riedesel and Fraser advocated a limited withdrawal, either to the line occupied briefly at Fish Kill when the army crossed the Hudson, or across the river to Batten Kill, and an attempt to restore the lines of communication. But Burgoyne again emphasised the inflexibility of his orders, which demanded, he felt, one more attempt to get to Albany.

The council, it would seem somewhat reluctantly, deferred to this decision to attack, but a compromise was reached whereby a reconnaissance in force of some 1,500 men should be sent round the enemy's left flank to observe the strength of the position, with a view to a full-scale attack on the following day. Should the reconnaissance find the position too strongly held, the army would withdraw to Fish Kill. If it was really intended as just a reconnaissance, it should have been carried out by far fewer troops, for a large force advancing towards the enemy's flank is liable to be misconstrued by them as a major attack to be dealt with by a similar, or preferably larger, body of troops. This is exactly what happened in what became known as the Battle of Bemis Heights. And, from the evidence Burgoyne gave later to Parliament, it would seem that it was exactly what he intended should happen.[17] It was a gambler's throw, for Gates was to fight the coming battle with 6,444 Continentals and 6,621 militia – more than double what Burgoyne could muster.

The reconnaissance force left the defensive position shortly before midday on 7 October. It had been carefully chosen for mobility and fire power – light infantry, grenadiers, jägers, heavy cannon and howitzers – and was divided into two wings and a centre. Fraser commanded the right wing, Riedesel the centre and Major John Acland of the 20th Regiment was in command of the left. Burgoyne and his staff rode with the centre. In the end, the total was nearer 2,000 than 1,500.

After marching for about three-quarters of an hour, the force halted in the area of Barber's Farm and from there 150 irregulars were dispatched to reconnoitre Gates's extreme left. Wilkinson, who had observed the troops at their halt, informed Gates that he was about to be attacked and, on this occasion, the American general was content to give battle. Acland's men on the left were the first to be attacked, but shortly afterwards Fraser's right wing was subjected to a withering fire. Both wings fought with tremendous zest, but they were heavily outnumbered and forced to give ground. Fraser, always prominent on his white horse, was an obvious target for an American marksman, and in the course of the fierce action he fell mortally wounded. Acland too was lost, being severely wounded and captured.

The withdrawal of both flanks left the centre very exposed. Here not more than 300 Germans faced some 3,000 Americans. The Germans fought with great gallantry, well supported by the Hesse-Hanau artillery but, by the middle of the afternoon, it was clear to Burgoyne (whose bullet-riddled coat testified to his front line courage) that the whole line must withdraw behind the fortifications. There they might have held out until Burgoyne was able to extricate them safely, had it not been for the élan and courage of Benedict Arnold. This unpredictable soldier but brilliant leader had quarrelled with Gates and lost his command, but once battle was joined he was out before anyone could stop him, rallying scattered units and taking charge of first one brigade and then another, bringing sense and enormous enthusiasm to the American battle.

He quickly saw it was useless to throw troops against the strongly defended Balcarres Redoubt (into which almost all the British had withdrawn), so he bypassed it and turned his attention to the much less strongly held Breymann Redoubt. The Germans there fought with their customary courage but, when their commander was killed, they abandoned the position. The remnants of Burgoyne's army were now threatened in flank and rear – but darkness came in time. The fight was broken off and the British withdrew to the area of the Great Redoubt. Their casualties had been grievous, Burgoyne having lost 176 men killed, 200 wounded and 240 captured out of about 2,000 actually engaged. The American casualties were scarcely more than 200.

On 8 October, Gates's men occupied the line previously held by the British and opened up a lively cannonade on the Great Redoubt. Burgoyne saw very clearly that his position was untenable and decided to fall back to Saratoga. It was 9pm when the first troops left the camp and the weary and dispirited army, their supplies almost exhausted, marched – together with the ladies, although the wounded had to be left behind – to their old positions at Fish Kill. The weather was appalling and so was the mud, the horses struggling fetlock deep and the guns sinking beyond their wheelhubs, but at least the Americans did not pursue. Gates had no reason to hurry, for the trap was virtually closed, and his main army did not leave their camp until the late morning of the 10th.

At Fish Kill, Burgoyne took up a strong defensive position on rising ground with good fields of fire. From here he sent a force of six companies to reconnoitre the crossing near Fort Edward and to repair the broken bridges over the streams leading to the Hudson. It was a fruitless errand. Stark had already blocked the crossing, and anyway the reconnaissance had to be recalled at the approach of the American army. Gates had mistaken the reconnaissance for the beginning of a further British withdrawal and had not seen the force return. He therefore thought he had no more than a rearguard to deal with and advanced boldly across the open, concealed only by a clearing mist which, most fortunately for him, held until after a deserter had told him that he was approaching the whole British army, strongly entrenched. It was Gates's lucky day and he knew it. There would be no further adventures of that kind, and he would now settle down to bombard the position and starve the British into surrender.

On 12 October, Burgoyne held a council of war. He offered his senior officers four choices: to stand and wait upon news from Clinton; to retreat with the guns; to withdraw by night without the guns or baggage; to skirt round Gates's extended left and march rapidly for Albany. The decision was to withdraw by night in great secrecy, but this had to be abandoned when scouts reported that secrecy was impossible for there were many enemy parties in the area, and that the western and eastern tracks up the Hudson were blocked. The next day Burgoyne called another council, and asked the assembled company if they thought an army of less than 5,000 facing an enemy of at least 12,000 men could honourably surrender, the answer was a unanimous 'yes', and under a flag of truce Gates was asked for terms.

Unaware of Clinton's exact whereabouts, Gates was anxious to have done with Burgoyne and, in the protracted discussions over several days, he agreed to exceptionally generous terms. Burgoyne rejected his original demand for unconditional surrender and insisted that his troops should march out of camp with the honours of war and be given free passage back to Britain on the understanding that they would not serve in America again during the war.* He also insisted that the terms agreed upon should be called a convention and not a capitulation. When eventually these and other terms were agreed the Convention of Saratoga was signed on 16 October. The next day Burgoyne, wearing full dress uniform, led 1,905 British and 1,594 Germans[18] out of camp to lay down their arms. Burgoyne surrendered his sword with the words, 'The fortune of war, General Gates, has made me your prisoner' to which Gates replied, 'I shall always be ready to bear testimony that it has not been through any fault of your Excellency',[19] and returned his sword. In victory Gates had behaved with moderation and magnanimity, in defeat Burgoyne had behaved with fortitude and dignity.

*   *   *

The Saratoga campaign was the turning point of the war. In American eyes it was to become a heroic apologue in the cause of freedom for now, with the aid of powerful allies, American fortunes would broaden through the years. On the other hand, with General Burgoyne's surrender, British aspirations, already sadly shaken, were doomed. The French were only waiting for such a signal, nor would the Spaniards be far behind. Soon a government hard pressed at home would have to contend with forces much superior to those that had been encountered, with no great measure of success, in the early stages of a colonial war.

If the original plan had been adhered to and properly executed it is just possible that with the line of the Hudson occupied, New England states isolated and a substantial American army defeated, the British could, with some sort of compromise peace, have extricated themselves from the imbroglio they had created. But the contemplation of historical 'ifs' is never very rewarding. The hard fact is that, through a series of mistakes and muddles in London and America, the campaign was a disaster. Even if it had been successful, nothing very spectacular would have been achieved, for there were not sufficient troops in America to keep New England isolated and to defeat the enemy armies. Howe had foreseen this all along, and constantly pleaded for more men. Such pleas would have been unnecessary had he taken the opportunities offered on Long and Manhattan Islands to destroy Washington's only army.

Great hopes had been entertained of the northern army, and when the news of Saratoga reached England on 2 December (nine days ahead of Burgoyne's

---

*Sadly, the blunders of the campaign did not end at Saratoga. Howe wasted many months insisting that the troops embark from a British port, and Burgoyne in a fit of exasperation used the unfortunate phrase 'the public faith was broke'. All this gave the casuists of Congress the chance to renege on some of the terms of the Convention, and the troops were to remain prisoners in America for the rest of the war.

official dispatch) there began inside and outside of Parliament inquisitions, accusations and recriminations. These tergiversations lasted well into 1779. Burgoyne eventually obtained his parole in April 1778, and arrived home in May. Germain refused him his undoubted right to a court-martial, but grudgingly allowed an inquiry in camera. Nevertheless, there was no lack of fuel for taunt and rejoinder, and the Opposition were in full cry over a long period of time for Germain's blood. Fox with his rational mind and forceful eloquence was most telling in debate and, together with Shelburne, Burke and a fast fading Chatham, sought to put the whole blame upon Germain, while exonerating Howe and Burgoyne.[20]

Germain was fighting for his political life. As already mentioned, no sooner had he heard of the Bennington troubles and suspected that blame for the campaign's failure might fall on him, than he started looking for a scapegoat. At first he settled on Howe, whom Burgoyne was also inclined to blame until Opposition members made it clear they would withdraw their support if he persisted. But it was not long before Germain switched his attack to Burgoyne, who he declared was solely responsible for the plan, and an enduring antagonism sprang up between the statesman and soldier.

The venom of the Opposition attacks and the damaging indictments of its leaders had some support in military circles as Carleton's letter of commiseration to Burgoyne in November 1777 shows, 'This unfortunate event, it is to be hoped, will in future prevent ministers from pretending to direct operations of war, in a country at 3,000 miles distance of which they have so little knowledge . . .'.[21] However, for the time being, Germain survived.

The principal blunder was undoubtedly his, although Howe certainly cannot be entirely exonerated. Nor indeed can Clinton, who in the long drawn out post-mortems to this sorry affair remained comparativley free of censure. It is quite possible that had he been more prompt and more resolute, Clinton could have been of some use to Burgoyne. The reluctance of the river pilots to proceed farther up river can hardly be accepted as a valid excuse, and we know of no other.

Burgoyne received a large measure of support throughout the long drawn out debates and confrontations, nevertheless it is permissible to ask whether Carleton, who so nearly got the command, could have got through to Albany. The answer is almost certainly 'No', but he probably would not have committed what was perhaps the worst of Burgoyne's mistakes – his failure to go back when he knew that Howe would not meet him, and that the state of his supplies was becoming desperate. Carleton had conveyed to Burgoyne the instructions from Germain, which allowed that general and St Leger a loophole should their situation require it and there were no orders from Howe. It is quite probable that Carleton, who was not such a proud man as Burgoyne, would have taken the sensible course of saving the army when its ultimate purpose had been vitiated through the absence of Howe.

Burgoyne did, of course, make other mistakes. He shared with others an underestimation of his enemy; at Bennington there was a lapse of judgement, and on 21 September he made what seems with hindsight his second greatest error of the campaign in not taking his last chance of a breakthrough. Nevertheless, of the three principals concerned in this unhappy affair, Burgoyne's reputation was the least dented. His courage in battle was unassailable and his many human qualities endeared him to all ranks, who gave him their trust throughout. He had followed what he thought was the path of duty: that it did not lead to the glory he so much desired was unfortunate, but at least he emerged with his honour secure,

CHAPTER 9

The French Connection

WHEN THE NEWS of Saratoga reached London, Parliament and the British people would have realised that in the course of three campaigns, their armies had been worsted in two and achieved nothing in the third. Now, with the increasing threat of French intervention, what had begun as a troublesome colonial rebellion was in danger of becoming a major conflict which could – and indeed by 1780 did – escalate into a world war. And yet on the very day (11 December 1777) that Burgoyne's Saratoga dispatch arrived, Parliament recessed for the Christmas holiday until 20 January.

There were those inside and outside Parliament who fulminated against the government's seeming insouciance, and were anxious that the American Congress should be offered concessions that might lead to a peaceful solution. The war was already crippling the Treasury, the services payroll was forever expanding and the national debt had risen by almost 50 per cent in the last three years.[1] Foremost among those anxious for action was Lord Chatham (ailing he undoubtedly was, but not yet senile, as some historians, notably Fortescue, assert), who, in an effective speech, deplored the intended recess. 'At so tremendous a season', he told the House, 'it does not become your lordships, the great hereditary council of the nation, to neglect your duty; to retire to your country seats for six weeks, in quest of joy and merriment, while the real state of public affairs calls for grief, mourning and lamentation.'[2]

In this speech he pressed for conciliation (although not for the renunciation of independence), and he had support from the Rockingham Whigs* and others in the House. Moreover, there were a number of service personnel who had either been reluctant to take up arms against the colonists or, in some cases, had sacrificed their careers rather than do so. Such men, and many others, while not anxious to take an active part in the American struggle would be fully behind the government in countering any French threat. Indeed, at this time a number of English and Scottish cities offered to raise battalions, a gesture that in some cases was eagerly accepted, for it helped the government in their dilemma of finding troops to continue the war. There had been considerable opposition, from the King among others, to sending more of what

---

*Government at this time was more by groups than by parties.

was called the Old Corps (the existing army) from England in the country's present 'Weak State'. Fox, speaking for two hours and 40 minutes to a crowded House on 2 February, ranged over the whole American story from 1774 to the present time. He pointed out that Britain had wasted lives and treasure to no purpose, and that nothing could be gained by force alone. Now, when 'there is the greatest reason to prepare for a foreign war' the army in Great Britain was 6,000 below the peacetime establishment, and therefore 'it would be madness to part with any more of the army' in a war that had become 'impracticable'. He moved that 'no more of the Old Corps should be sent out of the kingdom'.[3]

The motion was defeated by 94 votes, and so the government had a mandate to pursue the war to a satisfactory conclusion. However, with the near certainty of French intervention, a new strategy was under consideration. It was now clearly recognised that the war in America could not be won without many more troops and, as these were not readily available, the land war should be restricted to the defence of Canada and the Floridas and coastal raids from the few bases still in British hands, with the possibility of limited operations later in the South. The emphasis in future would be on a naval blockade, and the safeguarding of British possessions in the West Indies. The King himself had proposed something along these lines in a letter to Lord North of 31 January 1778.[4]

So much depended on what the French would do. Ever since the early 1760s, they had been waiting for the opportunity to avenge their defeat in the Seven Years War and, under the capable guidance of the Duc de Choiseul, they had built up a powerful navy. That competent statesman and his successors well understood the value of the American trade to Britain. When there was the threat of rebellion in the American colonies, the then Foreign Secretary, Comte de Vergennes, quickly realised the benefits that could accrue to France and, through his agents, he arranged a considerable amount of clandestine financial and military assistance to the Patriots. Moreover, in 1776, Vergennes persuaded his opposite number in Spain, the Marquis Grimaldi, to join in this opportunity to humble Britain by supplying her colonies with arms and money, although when later it came to armed intervention Spain was the first to hold back, for Grimaldi's successor, Count Florida Blanca, and his king, Charles III, were uneasy at the prospect of a republic on the border of their American possessions.

Vergennes did not have matters all his own way in Paris, for Louis XVI was vacillating, and his Comptroller of Finances, Baron Turgot, was vehemently opposed to conflict with Britain. However, the very subtle blandishments of Benjamin Franklin and his fellow Commissioners were reinforced by Turgot's fall, the news of Saratoga and the constant pressure from Frederick the Great, who calculated that war between France and Britain would enhance Prussia's security. Consequently, France decided to treat with America. On 6 February 1778, a Treaty of Amity and Commerce, and a defensive Treaty of Alliance were signed in Paris, and war between Britain and France became inevitable. In the event of war the treaty of alliance was to give America a free hand on the

North American mainland and in the Bermudas, while France reserved the British West Indies for herself.

During all this time, North and his Cabinet colleagues showed no sense of urgency. The Prime Minister seemed to be floundering, out of his depth, into a morass of uncertainty. Courses of action were no sooner considered than discarded. At one moment the plan was to stand fast in America, and send troops and the navy for operations in the West Indies; then it was decided to send more troops to America and prosecute the war there to the utmost; and then again, on 11 February, North told the House that he would 'on the 17th instant, propose a Plan of Conciliation with America'.[5] Such indecision was born of tiredness and a lack of confidence. North had offered to resign, but the King would have none of it, for in no circumstances would he entertain as his First Minister Chatham, who had been suggested as the man most likely to lift the country's affairs out of the catalepsy by which they had been smitten.

The concessions North was to announce offered the Americans almost everything short of independence, which was the one concession they were determined to have. The duty on tea was to be repealed, there were to be no taxes for revenue, laws which excluded New England fishermen from fishing in Newfoundland waters would be repealed, the Charter of Massachusetts Bay would be restored, and there would be a full pardon to all engaged in rebellion. Such were the main concessions that comprised this complete volte-face of a policy that had involved Britain in nearly three years of bloody struggle with her colonists, and which were soon to involve her in a major war with France.

A second bill was introduced authorising commissioners to be sent to America to inform Congress, the American Commander-in-Chief, and any other person of importance of these concessions, and to negotiate a peace. The three principal plenipotentiaries – George Johnstone, a rough-hewn ex-governor of Florida, William Eden, an MP, and the Earl of Carlisle, a recently reformed rake – did not leave England until 22 April, and arrived at Philadelphia at the very time General Clinton was preparing to abandon that city, an operation of which the government had failed to inform them. They had not been well chosen, but no body of men bringing such paltry offerings could have made any progress with Congress, which six months earlier had passed a resolution that no treaty could be made with Great Britain inconsistent with independence.[6]

In amazed silence the House heard North's speech introducing his bill. Fox was the first to rise, his face a mask of woeful virtue, for was it not a fact that the propositions just made 'did not materially differ from those which had been made by an honourable friend of his [Burke] about three years ago; that the very same arguments that had been used by the minority, and very nearly in the same words, were used by the noble lord upon this occasion.' He felt he could not refuse his assent, but asked if these concessions 'should be found ample enough, and should be found to come too late, what punishment would be sufficient for those who adjourned parliament, in order to make a proposition of concession, and then had neglected to do it until France had considered a treaty with the independent states of America?'[7]

The bill was passed without a division, but there were those in both Houses indignant at this sudden abandonment of the government's whole American policy. Not least among these was Lord George Germain, whose advice in the matter had not been sought by the Prime Minister, for even after Saratoga he was opposed to any concessions, and he now considered the proposals would make 'a dangerous peace'.[8] The King, who remained adamant on the question of independence, had given unqualified approval to North's concessions. It is difficult to be sure whether he genuinely believed in a change of policy, was weakening before a determined and strengthening opposition, or whether it was a case of *reculer pour mieux sauter* when the Americans threw out the proposals.

On 13 March 1778, less than a month after North had announced his package of concessions and more than a month before the royal commissioners would sail, the French ambassador in London informed the British Government of the treaties made by the King of France with 'the United States of North America'.[9] The declaration came as no surprise to the British Government, which would now consider that a state of war existed between France and Britain; ambassadors were immediately withdrawn, although it was not until the middle of June that the first act of overt hostility was to occur at sea. King Louis and his government would regret this desire for *revanche*. They had entered a war from which they would gain nothing, and which led to national bankruptcy that in turn led to the summoning first of the Assembly of Notables then to the Estates-General, and finally to the Revolution.

The effect on North of the French king's rescript was to further diminish his fortitude of spirit, and he again begged the King to accept his resignation and send for Chatham, 'for the present Ministry cannot continue a fortnight as it is'. There followed a series of letters between the King and North in which the former, although prepared to accept one or two members of the Opposition such as Shelburne, or even Chatham, 'to the support of your administration', would not in any circumstances entertain a Whig ministry. He told North 'no advantage to this country nor personal danger can ever make me address myself for assistance either to Lord Chatham or any branch of the Opposition honestly I would rather lose the Crown I now wear than bear the ignominy of possessing it under their shackles'.[10] North was persuaded to continue in office, but he was clearly not up to shouldering his huge burden, and the King was mightily relieved when on 11 May 'the Great Commoner', as Lord Chatham was called, died.

In spite of the earnest endeavours of Vergennes to bring both branches of the House of Bourbon into conflict with Britain, Spain held back for more than a year. Offers of territory in America were not sufficient to overcome king and government's fear for the safety of their existing possessions. Gibraltar would have been a different matter. That pawn was used as the price of their neutrality. When, eventually, Spain was persuaded into declaring war on Britain, she made certain that under the Treaty of Aranjuez France agreed to continue the fight until Gibraltar was returned to Spain. Spain's entry into the war on 21 June 1779 was to be followed by that of Holland 18 months later. In March of that same year, the Empress Catherine of Russia initiated the so-called Armed Neutrality, by which the ships of European maritime nations were to be free to carry merchandise, other than arms and ammunition, into any port including those of the belligerents. There was little that was neutral about it, for it was principally aimed at England, who, in the eyes of most foreign governments, was no longer attempting to quell a domestic rebellion but was engaged in a war of conquest.

It was necessary to look at these events of the future, for they concerned the position of the British navy when the war became very largely a naval affair. It has already been said that at the outbreak of the rebellion the state of the British navy in regard to ships and personnel left much to be desired. But strenuous efforts were immediately made to bring existing ships up to standard, commission new ones and, preferably not through press gangs, to recruit and train sailors competent to man them. There are two interesting documents in the royal archives, one in the King's handwriting, showing the names and gun-capacity of 27 new ships to be completed during 1778, and the other, written by Lord Sandwich on 15 March 1778, giving the number of ships in the home fleet. There were 55 ships of the line of which 40 were ready for immediate service, and 46 frigates and sloops as well as eight armed ships of 20 guns. A further eight ships of the line were awaiting crews to man them, and nine were in service in foreign parts.[11] These latter were mainly in American waters and the Howe brothers consistently, and rightly, complained that they were insufficient to support the army and maintain off-shore patrols.

When war broke out with France and later with Spain, the Royal Navy had immense responsibilities. There was a very real threat of invasion, Gibraltar was under siege and had to be supplied by sea, there were stores and troops to be convoyed across the Atlantic, and the West Indies had to be guarded against French attack. By 1778, the French had twenty-one 64-gun ships in American waters and a further 43 out of Toulon or Brest. Nor could the small American navy be completely discounted; it is true they had no ships of the line, but Washington had commissioned a number of frigates and sloops and, although most of these had been captured or destroyed, there were a large number of privateers that operated with success against British merchantmen, particularly in the Caribbean. The British navy could never be certain of supremacy in either theatre of war.

Meanwhile, in America, Sir William Howe had sailed for England in May 1778. Lord Amherst was asked to replace him as Commander-in-Chief but declined. Germain would not tolerate Carleton, Cornwallis was considered, but Clinton, although having caused difficulties in the past, was eventually chosen. He was a perfectly capable general without being brilliant, and he was put at a disadvantage straightaway by being ordered to keep the war going on the mainland without ever having sufficient troops to make a serious impression. Clinton learnt of his appointment in April, and on 8 May he went to Philadelphia to take over from Howe. Shortly afterwards, a frigate arrived bearing two dispatches from Germain dated 8 and 21 March.[12] The latter superseded that of the 8th, which had been drafted before the French declaration had been given to the government. These were the first of constant and often contradictory orders or recommendations that Clinton was to receive from Germain.

The dispatch of 21 March outlined the new government policy. In particular Clinton was instructed to evacuate Philadelphia and withdraw to New York. Should the peace commission fail, it might be necessary to abandon New York, holding only Rhode Island and Halifax, and sending the rest of his troops to Canada. He was to embark 5,000 men with artillery to attack St Lucia, and a further 3,000 were to go to Florida. And despite these sizeable detachments, he was expected to act offensively on the New England coast. The responsibility for these orders, with their curious amalgam of defence and attack, and which came in the form of an instruction from the King, 'given at our Court of St James's, the 21st day of March 1778', cannot be attributed entirely to Germain. They were a Cabinet decision and indeed emanated partly from the King in a letter to North of 17 March.[13] It is arguable that it was wrong to evacuate Philadelphia, and certainly Rhode Island could not be held without New York, on which it depended for fuel. Nor was there great merit in tying up new levies in Canada, for although the Americans would certainly have liked to have taken it they were not in a position to do so.

The dispatch, outlining the withdrawals, went on to say that the reinforcements of 10,000 to 12,000 men mentioned in the earlier dispatch would now be no more than three regiments originally destined for Nova Scotia.[14] It is small wonder that Clinton was bewildered and depressed by this inauspicious beginning to his new command. For a general who made a habit of tendering his resignation when in the least thwarted it is, perhaps, not surprising that six months later he unsuccessfully sought permission to be relieved of this 'mortifying command'.[15]

The British commissioners arrived in Philadelphia at the beginning of their ill-starred peace mission on 6 June 1778. They had fully expected to find General Clinton's army powerfully based in the American captial, thus enabling them to negotiate from a position of strength. Instead, they were greeted by the two Commanders-in-Chief (Lord Howe had been ordered to join the Commission) who shamefacedly told them that Philadelphia was about to be evacuated. For some inexplicable and unpardonable reason, the commissioners had never been informed of this, although the decision had been taken a month before they sailed, and during that time Eden had been working in close association with North and Germain. But for this precipitate retreat from Philadelphia (for that is what it was), at the time of their arrival,

Eden felt they might 'in two or three months have made a great impression' although Carlisle, writing privately later, was nearer the mark in saying 'our offers of peace wore too much the appearance of supplications for mercy from a vanquished, and exhausted State'.[16]

Their position was certainly embarrassing, but neither the time nor place of their arrival was likely to have made much difference to the outcome of their mission, for Congress, as its President said in answer to a letter of 9 June requesting a start to negotiations, would only be ready 'to enter upon a treaty of peace and commerce when the King of Great Britain shall demonstrate a sincere disposition for that purpose by an explicit acknowledgment of the independence of these states, or withdrawing his fleets and armies.'[17] These were terms which the commissioners had no mandate to negotiate. They sailed to New York in one of Howe's ships, where they continued to have no success and in July they asked Germain for leave to return, but he urged them to stay longer. In October they published a manifesto which offered peace terms to each state, and which very foolishly had a threatening peroration promising total destruction if their offers were not accepted. It is impossible to be certain, but this threat may well have been uttered at Germain's instigation. The manifesto, or proclamation, was greeted with derision by those few who troubled to read it. Even greater derision and discredit was the commissioners' lot when Johnstone was exposed in an attempt to bribe some Congressmen and others who he had thought could be bought. In November, the commissioners sailed home, having achieved nothing but greatly upset the Loyalists and uncommitted Americans.

When Clinton and Lord Howe were considering the evacuation of Philadelphia, it was decided to ignore the instructions to take the army by sea. There were a number of good reasons for going by land. New Castle, the port of embarkation, was some 40 miles distant; there were not sufficient transports to take the 3,000 Loyalists who could not be left to the mercy of the Patriots, nor could the many horses and stores accumulated by the army during their time in Philadelphia be shipped; there was also the question of New York's security, for Clinton commanded the principal army, and had it been delayed at sea through bad weather, there was little to stop Washington from taking that vital British base. Although unbeknown to Clinton at the time, the Comte d'Estaing's French squadron, which had left Toulon on 15 April, was drawing very close. Accordingly, the British, with an immensely long wagon train, left Philadelphia on the morning of 18 June. That evening, the American army entered the city.

Washington regarded the British retreat as a victory, which in a way it was, and with an army almost equal in numbers, he decided to give chase and, if possible, engage. He had had an extremely difficult winter in his tented camp at Valley Forge, where his soldiers suffered from acute shortages of food, clothing and money. Officers resigned, men deserted and, worst of all, a cabal, led by the Inspector-General (Thomas Conway), attempted to replace him as commander-in-chief by Gates. Washington could be thankful that General Howe, just a few miles away, had taken no advantage of his army's deplorable state. By the spring, matters had greatly improved. The cabal had been defeated and the consequential enhancement of Washington's popularity combined with an improvement in the weather greatly stimulated recruitment. Men joined from nearby and from New England, secure since Saratoga, and just at this time there arrived a man of the moment.

It is not exactly clear what tempted the Baron von Steuben, a notable soldier of fortune, to forsake the lucrative high-ranking positions he claimed to have held in the various princely courts of Europe (including that of Frederick the Great) in order to serve the dishevelled, undisciplined and untrained American army. But with enormous energy he had within six months transformed the troops into a competent fighting machine, and instilled the officers with that indefinable blend of qualities encompassed in the word 'leadership'.

The value of Steuben's training was very quickly evident in the action that took place on 28 June at Monmouth Courthouse against Clinton's retreating army. Clinton had crossed the Delaware at Gloucester Point and marched for Amboy but, on learning that Washington's army had crossed at Corell's Ferry and was marching parallel to him, and that his crossing of the Raritan might be disputed by Gates coming against him from the north, he changed direction at Allentown striking east for Sandy Hook. On the night of 27 June, the British were in the Monmouth-Freehold area, while Lee, recently exchanged from being a prisoner of war and now in command of the American vanguard, was some five miles north of the British with Washington and the main body three miles in his rear. Clinton had decided to alter his original line of march, whereby a right and left column marched on parallel tracks, and ordered Knyphausen with the right column, consisting of 10 battalions and all the baggage, to form the van, while the left column, led by Cornwallis (who had returned with the Commissioners from one of his visits to England), brought up the rear. At first light on the 28th, Knyphausen began his march and Clinton let his long winding trail of baggage get well clear before Cornwallis moved off at 10am.

Washington had ordered Lee's vanguard to move quickly across ground made difficult by three ravines and to engage the British rearguard until he could come up to administer the *coup de grâce*. And it was soon after Cornwallis had begun his march that he found his force virtually enveloped by three columns of the American vanguard. Cornwallis immediately changed front and attacked, driving Lee's forward troops back, but with no serious loss of ground until Lee, who for some reason had always been against any engagement, decided to order a general withdrawal, which became something of a shambles by the time the most westerly ravine had been crossed. At this point Washington arrived and upbraided Lee for flagrant disobedience of orders. His army was now in great jeopardy. There is nothing more demoralising for advancing troops than to be confronted by bewildered

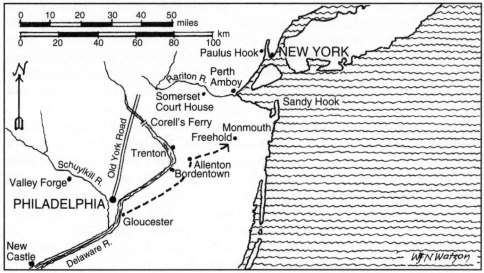

Clinton's Advance from Philadelphia to Monmouth Courthouse

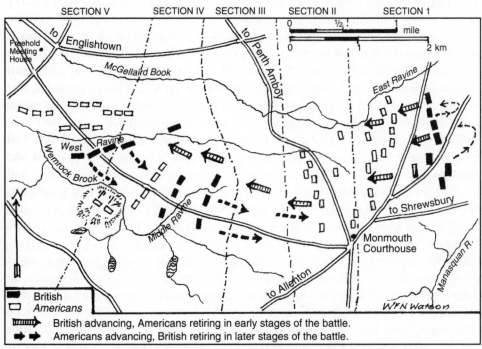

	British
	Americans

British advancing, Americans retiring in early stages of the battle.
Americans advancing, British retiring in later stages of the battle.

The Battle of Monmouth Courthouse, 28 June 1778

Right to Left

SECTION I	British proceeding eastwards; Americans (1st posn) attack rear.
	British face about and drive Americans westwards to...
SECTION II	American 2nd posn.; British continue to attack westwards across Sections I, II and III into...
SECTION IV	American 3rd posn; battle continues; Americans driven back over West Ravine into...
SECTION V	American 4th posn; rallied by Washington; British driven back eastwards towards East Ravine in Section II again; fighting ceases at nightfall; at 10pm British resume their march eastwards to the coast.

comrades in headlong retreat, but such had been the discipline inculcated into these men by Steuben, that Washington was able to steady his own force, and turn Lee's about to face their enemy again.

Thereafter the fight was hotly contested in every sense, for at midday the temperature in the shade of what few trees there were on the plain was 96 degrees Fahrenheit. Clinton had sent for reinforcements from Knyphausen but, even with them, Cornwallis's troops were driven hard and forced to fall back to the east ravine where night stopped the battle. At 10pm, Clinton resumed his march to the coast unhindered. Lee had given as his excuse for disobeying orders (for which he was found guilty at a court-martial) that he considered it unwise to bring on a general engagement against good troops. The Americans felt that his poltroonery had lost them the day, but Clinton is of the opinion that, having regard to the terrain, he was right not to attack and that, had he done so, 'his whole corps would probably have fallen into the power of the King's army'.[18] Casualties on both sides in this battle were light, possibly reflecting a lack of zest in the appalling heat.

Washington did not pursue but took his army up the Hudson to a camp in the area of North Castle from where he carried out a number of raids on British posts. Clinton and Lord Howe arrived in New York by land and sea respectively, just ahead of the Comte d'Estaing in command of 12 ships of the line, six frigates and a large number of troops. The French squadron was off the coast of Virginia on 5 July, the same day as that on which Clinton embarked his troops from Sandy Hook. Had he decided to go by sea from Philadelphia, the delay in embarkation might very well have enabled d'Estaing, with his larger and more heavily armed fleet, to give the army great discomfiture.

The fact that Howe's fleet would be inferior in both numbers and fire power to d'Estaing's had given Germain great anxiety and he had pressed the government to send reinforcements. A fortnight after d'Estaing had left Toulon, it seemed certain that he was destined for America but Lord Sandwich, well supported by Admiral Keppel, refused to allow 13 ships of the line to leave home waters until definite confirmation was available. Thereafter there were endless delays (some admittedly due to weather) until the relief squadron under Admiral Byron, who was to replace Lord Howe in command, eventually left Plymouth on 9 June. By then he was far behind d'Estaing and, owing to some very severe Atlantic storms, his fleet became scattered. Not until 26 August did he reach Halifax himself.

D'Estaing was just too late to catch Clinton's troops disembarking at New York and soon realised that his larger ships drew too much water for him to cross the sand bar that protected the harbour and Howe's fleet. In consultation with Washington, he therefore decided to sail for Rhode Island, where General Pigot had some 5,000 troops. There in co-operation with General Sullivan, who commanded at Providence, he would attempt to drive the British from the island. Arriving at Newport on 29 July, he forced the British to burn seven small warships to avoid capture, but thereafter he encountered difficulties.

Pigot, whom Clinton had reinforced, took the offensive in the north and messed up Sullivan to such an extent that it was 9 August before he was in a position to co-operate with d'Estaing. On that very day Howe's fleet, strengthened by two of Byron's stragglers, anchored off Point Judith and d'Estaing, who had found Sullivan surly and far from co-operative, re-embarked his troops and put to sea seeking battle.

For two days the fleets manoeuvred for the weather-gauge, Howe's first purpose being to draw the French away from Rhode Island. In this he was very successful and when he decided to give battle, his changes of course to enable him to attack with the wind showed him to be a better tactician than d'Estaing – who was, after all, a soldier by trade. But just as Howe had manoeuvred his ships into the right position for attack, the storm that had been brewing intensified to such an extent that the sea became mountainous and battle impossible. Isolated fierce actions were fought in chance encounters, but both fleets were scattered and suffered great damage in the storm, that lasted with undiminished intensity for almost three days. Eventually Howe's ships got safely into Sandy Hook, where all but two could be repaired. D'Estaing's had not fared so well. Having put back to Newport to explain that he could no longer assist in the siege (which Sullivan was soon to raise) he made for Boston to carry out extensive repairs in comparative safety, Here the French remained, not greatly enjoying an alliance that, with the Bostonians at least, was beginning to wear very thin, until on 4 November they sailed for the West Indies.

On land the war in the North was entering a quieter phase. In December 1778, Washington had established his headquarters at Middlebrook, and had troops up the Lower Hudson. During the summer of 1779, the opposing forces engaged in a number of offensive and defensive operations of a minor nature up the river from Paulus Hook on the Jersey shore to Stony Point and Verplant forts, as well as in Connecticut and Penobscot, on the Massachusetts coast. But before that, in response to the government's leaning towards a Southern campaign, for which ministers had again, and with better reasons than before, high hopes of loyalist co-operation, Clinton had sent Lieutenant Colonel Archibald Campbell of Macleans's Regiment with 3,500 troops to take Savannah.

It was to be a combined operation, with General Prevost advancing from St Augustine with a similar number of men and was entirely successful. Landing at the mouth of the river two days before Christmas 1778, Campbell's force was guided through the treacherous swamps by a Loyalist negro. They very soon routed General Robert Howe's smaller patriot force, taking the town, about 500 prisoners, and a large amount of cannon, powder and other stores. Shortly afterwards, Prevost arrived from Florida, and with the capture of Sunbury and Augusta it was not long before Georgia returned to its allegiance to the Crown under a royal governor. However, a number of Patriots took to the hills to fight in guerrilla bands another day.

At the same time (beginning of November 1778), as Campbell sailed for Savannah, Clinton complied with the other order contained in Germain's dispatch of 21 March and sent General Grant with between 5,000 and 6,000 men to take St Lucia. The West Indies islands depended to a great extent on America for their food. Ever since 1775, there had been considerable unrest in the British possessions in the Leeward and Windward Islands, and indeed in the Bahamas, stimulated by American agents. This unrest increased when the French entered the war. In September, the Marquis de Bouillé had had little difficulty taking Dominica and now, with d'Estaing's fleet arriving at Martinique at the same time as the fleet under Commodore Hotham bringing Grant's troops reached Barbados, these islands became an important theatre of war.

Admiral Barrington, Commander-in-Chief Leeward Islands, did not allow Grant's men to disembark but sailed with them at once for St Lucia. There they landed in the north of the island on 13 December and, with very little difficulty, took possession. However, d'Estaing, who had been greatly reinforced at Martinique, made the recovery of St Lucia his first objective and, with a superiority in men and ships, he was fully confident. But his first attempt to land in Cul de Sac Bay was foiled by the skilful disposition of Barrington's ships and the time lost going north to Anse de Choc enabled Grant to take up a strong defensive position from which he repelled three attacks. D'Estaing then withdrew from the island leaving 400 dead behind.[19]

It had been a beautifully executed combined operation by both British commanders. Grant was to compound his tactical prowess by showing sufficient sense to disregard Germain's instructions to spread his force thinly around the islands. Instead, he kept sufficient troops on St Lucia to ensure its retention and the ability to play an important part when, three years later, French dominance in these islands was overthrown. D'Estaing meanwhile, with his force in disarray, had withdrawn to Martinique. There Admiral Byron kept him bottled up for many months. However, when Byron had temporarily to remove his ships for convoy duties in June, d'Estaing was able to slip out and take St Vincent and Grenada, which the French then garrisoned in sufficient strength to withstand immediate attempts at recapture. In September 1779, d'Estaing took his force to assist in operations, shortly to be recounted, for the recovery of Savannah.

The year 1779 was not a happy one for General Clinton. He was a man doing a difficult job under trying conditions. He felt insecure in his position and was sensitive for his reputation. Moreover, there was trouble in his own headquarters, where he had quarrelled with his senior aide (Major Drummond) and his popular adjutant-general (Lord Rawdon) both of whom had departed. His relations with senior naval officers, except for Commodore Collier, with whom he got on splendidly, were often very stormy and he was constantly plagued by Lord George Germain. It is not surprising that twice during the year he sought leave to resign a command in which he was being frequently interfered with by unrealistic recommendations from Whitehall.

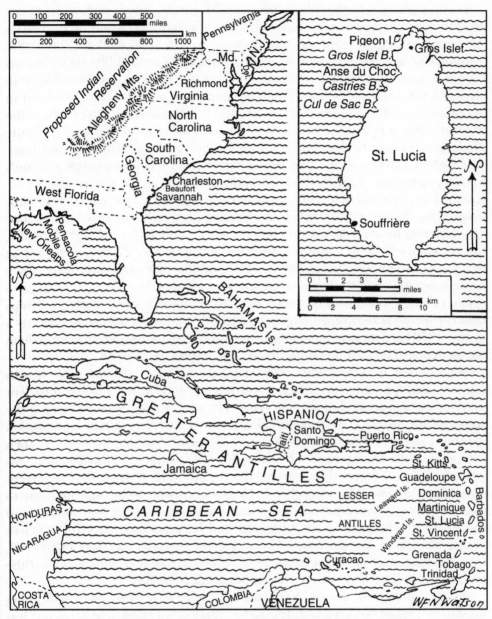

The Southern States and the West Indies, 1780

The expeditions to the West Indies and Georgia, which had taken some 10,000 men, had seriously reduced Clinton's fighting strength in North America and yet Germain was writing to urge him to 'a pitched battle with Washington' and, at the same time, to detach a force of 4,000 men for the Carolinas and, with similar numbers, to carry out raids on the New England coast.[20] Until reinforcements arrived, he simply had not enough men for these operations and still to protect New York. Much of the difficulty caused by these impracticable suggestions arose from Germain's habit of calculating the troops available to Clinton by their paper strength, whereas the Commander-in-Chief, very rightly, worked on the number of bayonets he had immediately available for any operation. This method Germain deplored, as he put in a letter to Knox, 'I do not like Clinton's dispatches as well as I had expected; he is magnifying the force of the rebels and limiting his own by the new-fangled way of computing his army by the number of rank and file fit for duty.'[21] In fact, in the summer of 1779, Clinton's effective strength was 27,000 men. Seven thousand of these were in the south, which was almost exactly a third less than the number shown in the Colonial Office return at that time.[22]

For nearly a year, Clinton had made no remonstrance to these constant, mainly ridiculous recommendations he had received from Germain. However, in May 1779 his patience ran out and, in a somewhat acerbic dispatch, he pulled no punches. He had been flattered, he wrote, on taking 'this difficult command' that he would have every latitude 'to act as the moment should require'. Instead there was a lack of confidence and he was mortified by the movements recommended for his army. He admits that 'your lordship only recommends. But by that recommendation you secure the right of blaming me if I should adopt other methods and fail.' He points out that he is on the spot with up-to-date intelligence. 'Why then, My Lord, without consulting me, will you adopt the ill-digested or interested suggestions of people who cannot be competent judges of the subject . . .?' He does not wish to be captious, but he implores the peccant minister, 'For God's sake, My Lord, if you wish that I should do anything, leave me to myself and let me adapt my efforts to the hourly change of circumstances. If not, tye me down to a certain point, and take the Risque of my want of success.'[23] Seldom can a commander-in-chief have written such a high-toned letter to his political master.

On 21 June 1779, rather less than a month after this letter had been dispatched, Spain declared war. Inevitably, this caused additional problems for the government and for the Navy, which by July had to contend with the serious threat of an invasion of Britain by a Franco-Spanish armada. This was at the instigation of the Spanish in the face of French scepticism, which was almost immediately justified when Spanish sloth in preparation, coupled with very rough weather, caused the French fleet to be at sea long enough for their crews to be decimated by scurvy. Nevertheless, Sir Charles Hardy, recently appointed to command the English fleet in place of Admiral Keppel, had only 38 ships of the line and a few frigates to oppose the allied armada of 66 ships of the line with transports for 30,000 troops. There was something near to panic in the Portsmouth and Plymouth areas where a landing was expected and where, through the usual muddles and mismanagement, there were only three militia regiments available at Portsmouth, and 4,800 regulars at Plymouth, a port of which the defences were hopelessly inadequate. For more than a month, great uncertainty prevailed, but early in September the sickness in the allied fleet had become so severe that it was forced to retire to Brest, remaining inactive there for several months.[24]

Fortunately, the invasion threat, although causing a deal of confusion and anxiety, proved something of a fiasco, but in the Caribbean the Spaniards had some success and, in the autumn of 1779, Prevost's troops were again in action in Savannah. Clinton, acting on one of Germain's recommendations, had dispatched a force of Germans under Brigadier John Campbell to strengthen the important posts of Mobile and Pensacola in West Florida, and to take New Orleans should the Spaniards enter the war. But the Spanish governor had earlier information of his government's declaration than did Campbell, which enabled him to capture all the small Mississippi posts and to forestall any surprise attack (which was Campbell's only hope with the inadequate force he had) on New Orleans.

In Georgia, it will be remembered General Prevost, in co-operation with Colonel Campbell, had taken Augusta at the beginning of the year. However, soon afterwards, being faced with a force of 1,500 men under General Ashe (General Lincoln's deputy commander) he had, for some reason that is not entirely clear, evacuated the town and marched south, pursued by Ashe. At Briar Creek on 3 March he turned on his pursuer and won a crushing victory. Augusta was reoccupied and Prevost thought himself strong enough to march on Charleston, summoning that town to surrender on 12 May. But terms could not be agreed and, finding the defences too strong, Prevost's force fell back on Savannah, hard pressed by Lincoln's troops.

Serious fighting then ceased in the hot months of mid-summer, until in September d'Estaing appeared off the coast with 22 ships of the line, 11 frigates and transports containing nine French regular regiments and two Irish regiments for the recapture of Savannah. Lincoln was on his way to join d'Estaing and Colonel Maitland with 800 British troops was hurrying from Beaufort to join Prevost. D'Estaing did not wait for Lincoln before summoning Prevost to surrender, but the latter delayed his refusal long enough for Maitland to march his 71st Highlanders across 50 miles of swamp and bog in an incredibly short time. By September, Savannah was invested, but Prevost had 3,700 men to defend it.

The normal pattern of siege warfare – sapping, sortie, and bombardment – was broken after a fortnight, for d'Estaing was anxious to have the job done before the autumn gales, and his 67 cannon were having little effect. He therefore decided that a major assault should be made by two columns, each of 4,000 men under himself and Lincoln, against the south-west and western defences, while American militia carried out a feint against the southern and eastern fronts. The assault went in on the morning of 9 October and Lincoln's column very soon got lost in the maze of tracks through the swamp. However d'Estaing's men reached the defences at the Springhill redoubt. For close on two hours, there was a tremendous fight in whch the French and Americans, although severely shattered by the British cannon, continued to storm the defences with outstanding bravery and, at one stage, had their colours flying from the parapet. But the effort needed to break through proved too great. D'Estaing ordered the retreat and the siege was raised. Maitland, whose men held the British right at the Springhill redoubt, immediately went in pursuit of the fleeing French and Americans, but a stout rearguard action, fought mainly by the volunteer regiment of Haiti Chasseurs, kept the British at bay.

British casualties had not been great, but the French lost 637 men killed and wounded in the siege, and the Americans 264.[25] Their greatest loss was Count Casimir Pulaski who commanded the Polish Legion. This gallant, charismatic cavalryman had come to America when the Polish cause in Europe was lost and, through his prowess on the field and charm off it, had won the friendship of Washington. After the battle, Lincoln withdrew to South Carolina and d'Estaing divided his fleet, sending part to the West Indies and taking the rest back to France. It had been a good beginning for the British to what was soon to become a major southern campaign.

It was not easy for Clinton to exercise control over his detached troops in the far south, but Cornwallis, whose wife had recently died, arrived from England as his second-in-command and this, together with the expected reinforcements, was obviously of considerable help. Such an experienced and efficient soldier as Cornwallis was a great asset, but Clinton had reservations. The two men had not got on well since the abortive Charleston expedition in January 1776 and the fighting round White Plains later that year. Moreover, Cornwallis had come with little more than half the reinforcements promised and with a dormant commission of succession, whenever it might suit the minister. This was not unusual. Indeed, Clinton had received a similar pledge when deputy to Howe, but it was a further blow to Clinton's confidence and a potent factor in the unhappy relationship that developed between the two men during the southern campaign.

Clinton fully shared the government's desire for this campaign, if for no other reason than that the security of Georgia needed the occupation of South Carolina. He had had it under consideration for some time, but a shortage of troops and the constant presence of a French fleet off the American coast had caused delays. The manpower situation was improved with the coming of reinforcements and, in September 1779, Clinton began a lengthy correspondence[26] with Admiral Arbuthnot (now his naval Commander-in-Chief), with whom he was never on very good terms, for the evacuation of the 4,000-strong garrison of Rhode Island. The Admiral had said that the island was no longer required for the navy, which was the principal reason for its original occupation, and the troops from the garrison would provide Clinton with a force just sufficient for him to embark an expedition against Charleston, and still have 10,000 men under General Knyphausen to safeguard New York. Accordingly, General Prescott (who had replaced Pigot in command) and his troops were brought out. The French were not slow to realise the importance of the island, and to occupy it with a strong force from which Clinton was unable to dislodge them.

* * *

By the time the news of Saratoga had reached London, the British people had endured at least three years of muddles, mistakes and unfulfilled expectations. The mother country had failed to bring her supposedly erring children back to their allegiance either by peaceful means or by limited force. Now there was a need for urgent action, either to prosecute the war to the utmost, or to come to sensible terms. Men were needed of commanding spirit and intellectual superiority, men of vitality and energy, qualities not easily discernible in the King's ministers, especially in his First Minister, who had lost control and who clung to policies he no longer had faith in. The King should have let North resign. His loyalty to his sovereign had become a liability.

In Parliament, talent was chiefly to be found on the Opposition benches; there were men there who could clearly see the evolving destiny of America, which stood out like some giant tree against the horizon of human history. Men who might have concluded an honourable settlement based on independence but not total severance and who might have avoided a swiftly approaching major war. But King George remained obdurate. He would not tolerate the Whigs and so a tired ministry had to blunder on.

The government's unwarranted procrastination in not offering concessions to the Americans until after they had signed a treaty with France was unlikely to have made much difference to the outcome of the commissioners' mission, but the delay in sending them and the failure to inform them of the evacuation of Philadelphia destroyed what slender chance of success they might have had. It should have been obvious to the most myopic minister that by then without independence nothing counted. It was a classic case of too little too late in a retreat from a bungled policy of former years. And whoever instigated the commissioners' inflammatory proclamation merely intensified the bitterness and passions of the hour.

It was impossible not to feel sorry for Clinton: when France entered the war he was doing a job that was too big for him. There should have been an independent military command in the Caribbean well before 1780, and of course there was a need for more troops and ships. These were not available because the armed forces were still suffering to a certain degree from that constantly recurring British failing of allowing them to be run down in times of peace. Commanders-in-chief are bound to be subjected to political directions in a democracy, but Clinton can justly claim that the orders he received from Germain, thinly veiled as recommendations and based on an improper understanding of local conditions, lacked perspicuity and caused confusion.

The dispatch of 21 March, admittedly a collective Cabinet decision and not a personal directive of Germain's, is nevertheless an example of this, with its contradicting attack and defence policy. In it, Clinton was ordered to abandon Philadelphia, the American capital that had been captured with such effort, and to dispatch a large body of troops to the West Indies. Philadelphia was not a good centre from which to conduct operations, and with the French in the war, some realignment of available forces was necessary. But there was a morale question to be considered and, if we are to believe Lord Carlisle, albeit recently arrived and with a vested interest, the troops there were strong enough and fit enough to have walked over Washington's army whenever they pleased. To evacuate was probably the right course, but there had been sufficient time to consult the newly appointed Commander-in-Chief, who was to have the responsibility for carrying out any political decision.

Most of the mistakes, and there were quite a few, that occurred at this time were political. Nevertheless, the service chiefs cannot escape censure for the evacuation of Rhode Island. Clinton was at pains to put the onus on Admiral Arbuthnot, who was supported by his commodore, George Collier, and by Lord Cornwallis. However, the ultimate decision was his. He had been told in the instruction of 21 March that Rhode Island was to be held, even if the rest of North America had to be abandoned – but he had also been urged to begin operations in the south. It was another example of two orders that were not compatible. Without Prescott's men, he could only go south by dangerously reducing the numbers to be left in New York. In retrospect, it is easy to agree with those who have said that the evacuation of Rhode Island was a blunder, but by then Clinton was in the unenviable position of having to measure, one against the other, the requirements of at least three campaigns in widely separated areas: a formidable task for a master strategist, but a truly daunting one for an average general lamentably short of troops.

CHAPTER 10

The Carolinas: 1780

THE COLONIAL SECRETARY had for some time pinned great hopes on a southern campaign. The government badly needed to show some outstanding success by the armed forces in America, for since the Saratoga disaster only one or two distinguished naval actions, and the taking of Georgia and St Lucia had relieved the Cimmerian gloom that shrouded British endeavours in North America and the West Indies. The success in Georgia had been a particular encouragement to Germain and greatly strengthened his desire to go south, for it had seriously affected the French trade with the southern colonies and the West Indies. He felt that an added bonus in the reduction of the Carolinas would be a further damage to French interests.[1] Hitherto, Clinton had been kept lamentably short of troops with which to fulfil his many commitments, but recent reinforcements and the evacuation of Rhode Island, had enabled him to raise a force that with local Loyalists should be sufficient for a successful attempt on Charleston and beyond.

The royal commissioners, on return from their frustrating mission, had been over optimistic in portraying the amount of untapped Loyalist support available in the colonies. To what extent their report included the southern colonies, where of course they had not been, is uncertain, but their enthusiasm lent strength to Germain's expectations from the south. Early in 1779, new proposals had been made for bettering the pay, conditions and provincial rank of loyalist soldiers, but the response had not been great, and the total strength of the provincial corps in that year represented a substantial decline in the rate of growth from that of the preceding year.[2] Nevertheless, the government was not easily distracted from pursuing the loyalist will-o'-the-wisp, largely because growing opposition in England to sending more and more troops to America helped ministers believe that a strong militia would satisfy Clinton's demands. It will be seen how wrong this thinking was, for militia behaviour in both armies throughout the war was totally unpredictable.

Clinton, together with Lord Cornwallis, sailed from New York with 7,600 men on 26 December 1779. The fleet ran into some appalling weather; huge waves, impelled by gale force winds, crashed down on the vulnerable transports in a swirling maelstrom of white foam. The ships were scattered and terribly damaged, almost all the horses perished, and much of the artillery was lost. It was not until the end of January 1780 that the Tybee River was reached. Here it was decided to disembark Brigadier General Paterson with all the cavalry (most of them dismounted) and 1,400 infantry to make a diversion against Augusta, while the rest of the force proceeded by sea for the attack on Charleston. The choice between the Edisto or Stono inlets as the best landing place was to exacerbate the already uneasy relationship between Clinton and Arbuthnot and set the pattern for the rest of the campaign. On this occasion, largely due to Captain Elphinstone, Arbuthnot was overruled and disembarkation began on Simmons Island at the Edisto inlet on 11 February.

To land at Simmons Island meant a difficult march across John's and James' Islands with bridges to be built over the larger waterways such as Wapoo Creek, and small parties of American militia to be dealt with. However, Clinton considered it preferable to the longer sea route and, as it happened, he was right, for the ships were spared another fearful hurricane. It was probably while the army was disembarking at Simmons Island and awaiting those transports dispersed by the gales, that General Robertson arrived in a ship bound for New York. Clinton took the opportunity to entrust him with a message to General Knyphausen, commanding New York, to send reinforcements under Lord Rawden as soon as possible. At about the same time, Clinton recalled Paterson from Georgia for he saw that his troops would be needed to guard the magazine at Wapoo Creek and to man line of communication posts until the army had crossed the Ashley river. That operation was accomplished unopposed on the night of 29 March at Drayton Hall, some 15 miles upstream from Charleston, and on the night of 1 April, the first parallel was begun 'within 800 yards of the rebel works'.[3] This was almost six weeks since the landing on Simmons Island, and Clinton has been criticised for taking so long to cover so short a distance. However, without greater knowledge of the state of the two islands at that time, and reliable information as to the extent of enemy interference, it is difficult to pass judgement.

It was now, at this early stage of the campaign, that the Clinton-Cornwallis relationship took a sharp downward turn on its course of self-mutilation. Clinton was ever hopeful that one of his frequent applications to resign his command would be successful and, to that end, he had taken care to consult Cornwallis on every important operational and administrative matter. But now Clinton came to believe that Cornwallis was openly hinting that he (Clinton) was soon to be superseded, with the result that many of the officers already regarded Cornwallis as the chief. There seems no reason to believe that Cornwallis had in fact been spreading any such rumours, although there may be some truth that when Clinton told Cornwallis he would not be resigning 'His Lordship's carriage toward me immediately changed. And from this period he was pleased to withdraw his counsels and to confine himself to his routine of duty in the line, without honouring headquarters with his presence oftener than that required.'[4] For Cornwallis was undoubtedly very hopeful of getting the command.

John André, who was Clinton's highly regarded Personal Assistant, suggested that this apparent, boorish behaviour on Cornwallis's part was due to his desire 'to be quit of his liability'. But this can be only partly true and there was probably an element of pique in it. Sadly, this deterioration of relationship, which stretched back almost three years became worse as the campaign progressed. Inevitably there were divided loyalties where, in a well-ordered chain of command, there should be perfect trust between commander-in-chief, his staff and subordinate generals. There were occasions when Cornwallis seemed to forget that delegation does not mean surrendering personal command, and others when Clinton, from a distance of 1,000 miles, was tempted to interfere unnecessarily.

The American troops at Charleston were commanded by General Lincoln, whose approach at the head of 5,600 men prior to the siege of Savannah had undoubtedly helped to convince Prevost that Charleston could not be taken by him. Now, at the beginning of 1780, just a short time before the ring had closed round the town, Washington sent reinforcements and Lincoln had some 7,000 men there and in two outlying detachments. The seaward defences rested principally on the harbour bar and two forts. Fort Johnson, at the entrance to the harbour, which had been taken by Clinton on his march from the Edisto; and Fort Moultrie on Sullivan's Island. According to Tarleton, the Americans 'had a considerable marine force in Charles-town harbour' and later some of these ships were sunk to block the entrance to the Cooper river between the town and a small island called Shute's Folly.[5] Tarleton was at some distance from the action at this time, and his account must be suspect, for Clinton was to produce correspondence to show that throughout April he had repeatedly urged the Admiral to sail up the Cooper, and that Arbuthnot repeatedly procrastinated, but not because the river was blocked.[6] On the landward side, the defences stretched from the Ashley to the Cooper rivers, and were composed of a chain of redoubts and entrenchments covered by 80 pieces of cannon, with each flank being protected by a deep morass.

Clinton's traumatic sea journey and lengthy approach march had given Lincoln plenty of time to prepare these formidable landward and coastal defences, but he was never happy at the thought of being trapped by a force equal to his own in a city not easily defended from a combined land and sea attack. Lincoln's instinct, which he should have obeyed, was to evacuate his army and fight up-country, but the South Carolina government wanted the place held, and he thought that General Huger's three regiments of cavalry and some militia would be strong enough to keep open an escape route north of the town. However, a detached force of 1,400 men under Colonel Webster was soon to disillusion him. Webster reached Goose Creek on 12 April, and from there sent forward the British Legion under Banistre Tarleton which surprised and routed Huger's militiamen on the Cooper river near Monck's Corner.

Tarleton was to play a most important part in the forthcoming southern campaign, and a word about this colourful young officer would not be out of

place. Originally a cavalryman (1st Dragoon Guards) he had left his regiment to
volunteer for America, and had seen service under Clinton at Charleston in
May 1776, and under Howe throughout the latter's New York campaign,
serving on the staff and with regiments of horse before being given a captaincy
in the 78th Foot. In August 1778, at the age of 24, he was given command of
the British Legion which was a mixed force of cavalry and light infantry. It was
in this capacity that he gained his reputation, for he had most of the attributes
necessary for a successful leader of an independent raiding force – courage,
élan, a willingness to take risks, and the quality of decision to strike and destroy.
His proud and predatory men, whom he trained and led with great skill,
performed many notable exploits, but sadly all too often these were shamed by
the alloy of extreme cruelty, for discipline was not the Legion's strong suit. In
due course, Tarleton was to close a distinguished military career as General Sir
Banistre Tarleton Bt , GCB.

Meanwhile, in his fight at Monck's Corner, which cost him only one officer
and two men wounded, his troops captured a number of militia, a large
quantity of stores and, most importantly, enough fine horses to mount all of his
dragoons. This action enabled Webster to cross the Cooper and operate into
the Cooper-Wando area, where his and other troops would soon completely
close Lincoln's landward door. Some days earlier Admiral Arbuthnot had
ordered the heavy guns to be dismantled, and provisions and water removed

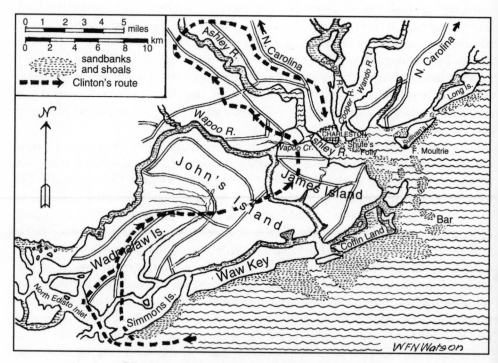

Clinton's Advance on Charleston, 1780

from the line-of-battle ships so that they could cross the bar. These were conveyed by a shore party under Captain Elphinstone to the magazines, and the Admiral, displaying considerable skill and courage, then sailed his ships past Fort Moultrie's guns with the loss of just 27 men killed and wounded, and anchored them within range of the shore. What American shipping there was in the harbour seems to have offered little resistance, and so now with Webster's men astride the northern exits, and Arbuthnot's ships in control of the harbour, Lincoln's force was truly bottled up.

On 19 April, Clinton received the reinforcements from New York which numbered 2,566 men and included the 42nd Regiment and Colonel Simcoe's Queen's Rangers. Simcoe had been given command of the Rangers in October 1777 and was to lead this very distinguished provincial corps with great distinction for the remainder of the war. Their 11 companies included a grenadier, light and Highland one, and they were a fine example of what properly trained and resolutely led Loyalists could achieve.

The arrival of these reinforcements enabled Clinton to send Cornwallis with 1,900 men to take over command from Webster on the other side of the Cooper. The situation between the two generals was becoming intolerable; there were arguments on almost every military subject, with Cornwallis at one point openly accusing Clinton of failing to enforce a ministerial order. Clinton had also become resentful at what he thought was Cornwallis's collaboration and tergiversation with Arbuthnot, whom he considered to be old, disagreeable and incompetent. It was André who sensibly suggested to Clinton that Cornwallis would be better away from the main siege in a semi-independent role. As we shall see, he was to perform this admirably and his closing of the escape route, in conjunction with Arbuthnot and Tarleton, was a considerable contribution to the success of the operation – a fact that Clinton was not prepared to admit.

Meanwhile, the siege of Charleston proceeded along perfectly orthodox lines with constant and heavy bombardment by both sides, and the occasional sortie by the besieged in an attempt to wreck British battery positions and the construction of the second and third parallels. Lieutenant Ewald, whose detailed daily account of the action is unrivalled, while highly praising the courage of the besiegers is very critical of British siege tactics.

Recalling an interesting conversation with his friend James Moncrieff, the engineer in charge of the siege, Ewald reflects on Clinton's generalship: 'I believe that he [Clinton] will reach his objective, provided no human life shall be spared. But I surely believe, for all his courage and tireless zeal, he would come off badly against any other army in Europe, and that he would not capture a dovecot in a European war'. And again, despite their friendship and Moncrieff's experience in other American sieges during the Seven Years War, Ewald writes of him: 'I assert once more that this man could hardly serve as an errand boy for an engineer in a European War.'[7] Thus speaks one brought up in the great tradition of the Marquis de Vauban.

However, despite Moncrieff's alleged inefficiencies, the siege was brought to a successful conclusion without coming to an assault. The breaching batteries had been brought up to the third parallel by 7 May (on which day Fort Moultrie surrendered to a storming party of sailors and marines), and four days later the howitzers were ordered to fire into the city with red-hot shot. The resulting destruction was sufficient for Lincoln to beat the chamade on the afternoon of 11 May. Seven generals were captured, and the British list gave a total of 5,677 prisoners including sailors from 20 captured ships (schooners and frigates). Casualties on both sides were light, the British lost 76 killed and 189 wounded (which included 84 Germans), and the Americans lost 239 killed and wounded, and 20 civilians killed in the bombardment.[8] Clinton could report to London a commendable victory, in achieving which he had shown considerable skill. It was to be the largest American surrender of the whole war (and indeed the largest for very many years to come), but its principal significance was local. Charleston was necessary as a base for future operations in South Carolina, but what little chance there ever had been of the British winning the war on land had long since gone. Now, everthing depended on command of the sea.

It had been agreed in London that on the taking of Charleston, Clinton and Arbuthnot should act as 'His Majesty's Commissioners for restoring Peace', but they had been given virtually no guidance, directions or freedom to negotiate. And not surprisingly, Clinton was soon to be complaining of this and that he could not work with Arbuthnot.[9] As a result, the restoration of civil government in Charleston made painful progress. Clinton's appreciation of the situation after the siege was far too optimistic; he envisaged little difficulty in mopping up the remaining militia and guerrilla bands in the province and then going forward to North Carolina. On 25 May, he wrote to his friend General Phillips 'I am clear in opinion that the Carolinas have been conquered in Charleston'.[10] It was left to Cornwallis to prove him wrong.

That general was to set off very soon after the capitulation at the head of one of three columns which the commander-in-chief sent out to pacify the backcountry and to encourage Loyalists to support actively the British forces. Cornwallis was to march up the north bank of the Santee to Camden, Colonel Balfour was to move to the south-west of that river to Ninety-Six and a third corps went up the Savannah River to Augusta. Meanwhile, Clinton remained in Charleston attending to administrative matters and attempting to enlist militia. For the latter purpose, he appointed Major Ferguson as Inspector of Militia.[11] Ferguson was exceptionally intelligent, and at the same time a man of action who thrived on the adrenalin of risks. At Brandywine, where it will be remembered he had spared Washington's life, he had lost the use of his right arm, but that did not diminish his relish for a fight and he much preferred the field to the routine work of an inspectorate.

While Ferguson went to Ninety-Six and joined Balfour, who considered him dangerously impetuous, Cornwallis found excellent use for that other notable Myrmidon, Banistre Tarleton. Colonel Buford, at the head of a Continental

The Carolinas and Virginia, 1780

regiment, was on his way to Charleston when it capitulated and had now fallen back beyond Camden. Tarleton caught up with him at Waxhaws on 30 May after an incredible march of 105 miles in 54 hours. In the subsequent engagement, the Americans misjudged the speed of the charging dragoons, and held their fire too long, with the result that most of Buford's 350 men were killed or captured, although he himself escaped. Tarleton's losses were only two officers and three privates killed, and 14 men wounded.[12]

The rout of these Continentals seemed to Clinton to remove the last barrier to a peaceful occupation of South Carolina and a successful invasion of the northern province. This being so, he prepared to sail for New York, where news of an approaching French fleet required his presence. Before leaving he issued a proclamation on 3 June. This had been preceded by two others, and some handbills, all of which appeared to have had a slightly beneficial effect on militia recruitment; but the wording of this latest publication, whereby the freeing of all prisoners from their paroles was tempered by a declaration that unless they returned to their allegiance they would be treated as rebels, proved most damaging. It was mainly through this and the cruelties perpetrated by Tarleton's men and other provincials, that peaceful citizens were stirred to hatred against the occupying power and greatly tempted to join the ranks of the redoubtable American guerrilla leaders, such as Francis Marion and Thomas Sumter in what had virtually become a civil war in the province.

Clinton sailed for New York on 8 June taking with him some 4.500 men, including nearly all the cavalry. Before he left, he issued detailed instructions to Cornwallis in which he gave him very wide powers, but too few men to carry out his far ranging commitments. His orders were to pacify the Carolinas, to hold a watching brief over Florida and Georgia and to be prepared to co-operate with Clinton in an assault on Virginia via the Chesapeake in the autumn, if the coast was free of French ships. For this purpose he had a total of around 4,000 British, German and provincial troops fit for duty, including six under-strength British regiments, the same number of provincials, and a small artillery component, but virtually no cavalry. Georgia and Florida had some militia, but Cornwallis had to find garrison troops in South Carolina, which left him with no more that 3,000 men for campaigning.[13]

On Clinton's departure, Cornwallis returned to Charleston to attend to administrative matters. There was much to be done in difficult circumstances and Cornwallis was quite capable of doing it; garrisons and their commanders had to be allocated and supplied, and kept in communication. Stores of all kinds arrived very spasmodically from Britain, usually via New York, distances were great and Clinton had taken almost all the wagons when he left. Feeding the army on local supplies was a problem, for Cornwallis forbade looting and the need to improvise transportation by land and water was equally difficult. Confiscated rebel estates went some way to solving Cornwallis's difficulties and, in this connection, he appointed Charles Stedman, a dedicated Loyalist from Pennsylvania (whose cogent views we have already noted in Chapters 6 and 7), to a senior post, in which he not only performed most expertly, but gained much interesting material for the absorbing two-volume history of the war he later wrote.

During this time the 25-year-old Lord Rawdon commanded in the field, centred on Camden.* He had outlying posts at various points on the frontier, and a substantial force under Lieutenant Colonel Cruger (who had replaced Balfour) at Ninety-Six. These detachments acted as sounding boards for troop movements in North Carolina, and enabled Rawdon to have advance warning of the approach of a large American force comprising the Maryland brigades, the Delaware regiment, and militia under Baron de Kalb. Rawdon, who had taken his troops forward some 14 miles north of Camden now ordered Cruger to send him four light companies, and recalled some frontier posts. These included some Loyalist and Legion troops at Hanging Rock, who had recently suffered a number of casualties while successfully withstanding a strong attack by Colonel Sumter's irregulars. On 9 August, Cornwallis received Rawdon's message of the American approach, and he left Charleston the next day, arriving at Camden on 13 August.

Rawdon's information estimated the Americans, now commanded by General Gates who had succeeded de Kalb, to be 5,000 strong (which proved an exaggeration), and Cornwallis had to decide whether to withdraw the army to Charleston and thereby lose South Carolina, or to stay to fight a numerically superior army and risk a major defeat. From what we know of Cornwallis it could not have taken him long to decide on battle.

Gates, who as a result of Saratoga had gained a spurious reputation for good generalship, had been appointed by Congress in preference to Washington's urgent request for Nathaniel Greene. In a short, disastrous campaign, he was to commit almost every mistake in the book from the time he chose the wrong route from Hillsborough to his hasty departure from the field in advance of his stricken army. Admittedly he had taken over an army of poor quality, lacking in arms, food, equipment and medical stores, but his leadership not only did nothing to help, but eliminated what slender chances of victory in South Carolina the Americans ever had.

On arriving at Rugely Mills, Gates decided against the sudden attack, for which he had marched his army over the shortest, but toughest and most dangerous route. Nor did he take the opportunity to outflank Rawdon before that general, fearing just such a manoeuvre, withdrew his force closer to Camden. Instead, from 13 to 15 August he did nothing. On the 14th, he was reinforced by 700 Virginia militia, which partly offset the sizeable detachment (which unwisely included 100 good Maryland Continentals)[14] that he had sent to Sumter, who was harassing the British line of communications. Gates's force now totalled 3,300 men of whom less than 1.400 were Continentals, the rest

*Rawdon later became Marquess of Hastings and, like his Chief, Cornwallis, served with great distinction in India.

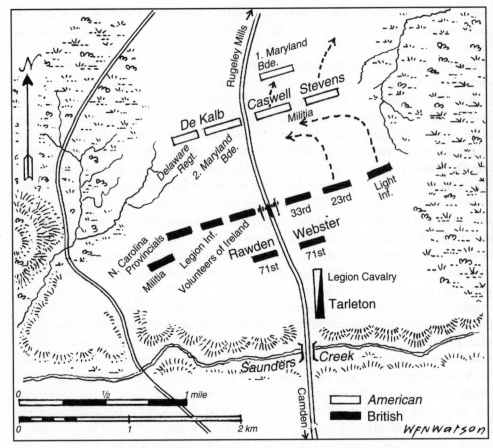

The Battle of Camden, 16 August 1780

unreliable militia, and at 10pm on the night of 15 August he moved them down the road to Camden. (A much larger number was attributed to Gates's army until accurate figures later became available. Cornwallis, in his dispatch to Germain of 21 August giving an account of the battle states the Americans were 'upwards of Five thousand men exclusive of General Sumter's detachment'.[15]) By a strange coincidence, Cornwallis began his march towards Rugely Mills at exactly the same time. He commanded 2,043 men, of whom only 817 were British regulars, but he expected the 844 provincials to fight well enough alongside the British, although the remaining 382 militia could not be relied upon.

After approximately four hours' marching, Tarleton's Legion cavalry and the advance guard of Webster's division bumped into the leading American troops at 2am north of Saunders Creek, which Cornwallis correctly gives as nine miles from Camden (Tarleton says only five).[16] When both parties had recovered from their initial surprise, some desultory firing took place for about half an hour with no damage done to either side. This gave Cornwallis the opportunity to learn something of the lie of the land from local guides. On being told that the road ran between two swamps, he decided the place would suit his purpose well and orders were issued for battle positions.

The British line extended across the road with flanks resting on the swamps. The site of the British and American positions is now densely wooded, but there is ample contemporary evidence to show that at the time of the battle this gently undulating ground, sloping steadily from north to south, was comparatively open – aptly termed by Cornwallis 'woody'. Colonel Webster had command of the right, with three light companies, the 23rd and 33rd regiments; on the left, Lord Rawdon had the Irish Volunteers, Tarleton's infantry, the North Carolina provincials and some militia; two six-pounders and two three-pounders under Lieutenant Macleod were in the centre of the first line. In reserve were the 71st and Tarleton's cavalry. Gates also drew up his army in two lines with his North Carolina, and recently joined Virginia, regiments under Colonel Caswell and General Stevens respectively on his left, and the greatly superior 2nd Maryland Brigade and Delaware Regiment across the road on his right under de Kalb; his light guns were divided among battalions, and he placed the 1st Maryland Brigade in reserve.

At first light, Gates, who earlier had contemplated withdrawing to avoid battle, was evidently dissatisfied with the dispositions on his left, and attempted to alter his line. Tarleton, who was sharply critical of Gates's whole performance, lists this as one of his many battle-losing moves, although there is some evidence that Stevens altered the American line when he saw the leading British troops deployed from parallel columns into line. Cornwallis was not the general to miss opportunities. As soon as he saw the milling chaos of Stevens's and Caswell's militia, he directed Webster to go straight at them with full force. The Virginians were appalled as swift destruction rushed upon them. They made no attempt to respond to Stevens's rallying calls, but throwing away their arms they bolted for the refuge of swamp and woodland. Webster was too good a commander to pursue, but swung his men round on the North Carolinians.

By now the heat and haze, coupled with clouds of reeking smoke, made for poor visibility, but what Gates's North Carolinians could see they did not like and very soon they were swelling the broad stream of fugitives struggling through the bog. Gates now had only de Kalb's Continentals on his right, and here the story was very different. Rawdon had opened his attack at the same time as Webster's but he was up against men who fought with stubborn courage, and more than once they were almost through the British ranks. But latterly the odds were too great, for with the American left gone, Cornwallis had sufficient troops to envelop them and, by the time he had sent Tarleton's cavalry to take them in rear, they were outfought and outmanoeuvred. They had performed with great valour, and none more so than their leader Baron de

Kalb; when many of his men decided at last it was time to go he stayed on the field until cut down, mortally wounded.

Gates and his officers could do nothing to halt the rout, his army was shattered and scattered. The carnage on the field and for miles around was very terrible, for Tarleton's men had little pity on the fugitives whom they pursued for many miles. Gates was back in Hillsborough in the quick time of three days, but less than 800 of his men rallied there a week later. He had lost, according to Cornwallis, some 800 men killed and 1,000 made prisoner, but the figure of 800 probably covers wounded. In addition, his camp equipage and all his cannon were taken.[17] Cornwallis lost 325 men of whom 69 were killed, 245 wounded and 11 were missing.[18] Charleston and Camden were the foremost British victories of the war.

It had been a straightforward, fairly simple set piece battle, brilliantly conducted by Cornwallis against a numerically superior opponent. Gates has been quite rightly castigated for his handling of the whole campaign, for he made a number of serious errors. But it has to be remembered that however competent or incompetent a general's strategy or tactics may be, it is the men at the sharp end who ultimately decide between victory and defeat. On this occasion, the completely unwarrated departure of half Gates's force early in the day was one of those incalculable hazards of the battlefield capable of overturning even the best laid plans.

The main American army had been beaten, discredited and thrown out of the province, but Sumter, an experienced guerrilla leader was still in the field with some 100 Continentals, 700 militia and two cannon. Cornwallis correctly guessed that he would now strike north towards the frontier, and he made arrangements to head him off by sending Tarleton in pursuit. The latter set off with 350 men and one cannon on the morning of 17 August, and eventually caught up with Sumter just north of Fishing Creek on the Catawba River. By then Tarleton's force consisted of only 100 dragoons and 60 foot soldiers. The rest had been left at Fishing Creek, too exhausted to march farther. However, the disparity of numbers did not matter, for the renowned guerrilla leader was most uncharacteristically surprised, quite literally, with his trousers down. Nevertheless, his troops put up what fight they could before being overwhelmed and suffering 150 men killed or wounded, and some 500 captured as well as two guns, and a quantity of arms and baggage. Tarleton's losses were only 15 killed and wounded.[19] It was a fine example of how a small force with the benefit of surprise can rout a much larger one. And also how a determined guerrilla, although only half dressed, can elude his enemy, for Sumter galloped off to safety!

The battle of Camden settled nothing, and the importance of the victory has often been exaggerated. Certainly it opened the way, so far as an opposing army was concerned, for an advance into North Carolina, but everything in the future would depend upon loyalist support, both active and passive. This was not forthcoming, for if anything Camden increased loyalist opposition. Besides

Sumter and Marion there were at least three other capable guerrila leaders still operating in South Carolina, and many flocked to their banners. The area between the Peedee and Santee rivers was, in modern parlance, virtually a no-go area for small military parties. Indeed, on one occasion a whole convoy escorting prisoners from Camden was ambushed and the prisoners were set free. Cornwallis took steps to disarm the roaming rebels there, and to hang those militiamen and others who had sworn allegiance to the Crown and were now in arms against it,[20] but the countryside remained extremely unsettled.

For some time, Germain had been anxious for an immediate occupation of North Carolina. London could never grasp the absolute unreliability of loyalist support, and Germain in particular had come to believe only that which he wanted to believe. Cornwallis was as anxious as anyone to cross the border, but he was realistic enough to appreciate the difficulties. Directly after Camden he was faced with a very large number of men either already sick or, in the appalling conditions of heat and meagre rations, rapidly becoming so. One of the major causes of British troubles, both then and generally, was the lamentable lack of proper medical supplies. He was short of almost everything, including wagons to transport what he had, and he was ever mindful of Clinton's admonition to safeguard Charleston at all costs. All these problems, he pointed out to Clinton, while impressing upon him the need to send a diversionary expedition to the Chesapeake, or else reinforcements to the Carolinas.[21]

Perhaps Cornwallis was tired of wrestling with endless economic and internal problems for which temperamentally he was not entirely suited; more probably, he realised that North Carolina must be subdued before the Carolinas were settled and that until then, the war could not be taken into Virginia; for whichever reason, he decided to march without waiting for any reinforcements, or indeed for the recovery of many of the sick. He left Camden on 8 September at the head of 2,200 men, all he could muster after leaving garrisons in Camden, Charleston, Ninety-Six and some border posts. His march was slow owing to the presence of so many sick men and the difficulty of obtaining supplies in a hostile part of the country. It was over a fornight before the army crossed the border and entered Charlotte.

While the march was in progress, Ferguson, now a Colonel, had been carrying out his largely unsuccessful attempts to enlist recruits along the north-west boundary of the colony close to the Blue Ridge mountains. He had with him 1,125 Loyalists – mostly militiamen – in whom he had the utmost trust. Never able to resist the chance of battle, he overreached his orders to keep in reasonable contact with Cornwallis when he heard that the American Colonel Clarke was in retreat after being repulsed at Augusta in Georgia. Ferguson was in the neighbourhood of Gilbert Town when he became aware that he was being hunted by no less than 3,000 backwoodsmen. He was a long way from base, hopelessly outnumbered and, as he was soon to learn, outclassed in the type of warfare in which these men excelled. He sent messages for help both to Cornwallis and to Cruger at Ninety-Six, none of which got through. Meanwhile, he fell back until being pressed so hard that he was forced to stand at bay.

His opponents had come together, and across Ferguson, more or less by accident. They were independent roving bands, each under its own commander and coming from North Carolina, Virginia and west of the Allegheny Mountains. They had three things in common, their determination to guard their various frontiers, their expertise with a rifle, and their skill in fieldcraft. As soon as they got scent of their quarry, these wild tramontane men got together and elected as their leader Colonel William Campbell, who commanded 400 men from Washington County, Virginia. They selected about 1,000 whom they mounted on their swiftest horses, and pursued Ferguson's force unremittingly until, on 9 October, they reached the feature known as King's Mountain, which Ferguson had decided he should be able to hold until relief arrived.

King's Mountain, 1,019 feet high, is one of several peaks on a six mile long ridge. Its rugged slope was strewn with boulders on ground shaded by numerous trees – mainly hardwoods with just a few pines; on the north-east side, the densely covered precipitous hillside made entry to the mountain from that direction nearly impossible. Ferguson realised he could be attacked from almost every side and he positioned his troops in an oblong position along the top of the hill. Against average troops he was probably right in thinking his position to be impregnable, but these backwoodsmen, as hardy as the Indians they had frequently fought and, like them, as agile and wily as mountain foxes, were no ordinary opponents. Tethering their horses at the base of the hill, they took up position on three sides and each contingent, under its own leader, swarmed up the mountain in a simultaneous assault.

Ferguson's men, especially those from Ninety-Six, did not betray his high encomium. For a seemingly endless hour, they displayed great courage, successfully pushing back frontal attacks with the bayonet, but they were up against a pack of rugged fighters who knew how to use cover, understood infiltration, and were deadly shots. Hemmed in on three sides, the Loyalists' position became desperate and Ferguson, noticing that some of his men were running short of ammunition, and others were faltering, rallied those he could with the shrill call of his silver whistle and vainly sought to break out. But it was a forlorn effort. His men were gradually pressed towards the north-east end of the crest, where they had the precipitous descent at their backs. Here they made their last stand and here Ferguson, conspicuous in his coloured hunting shirt, and fearlessly exposing himself on his agile charger, eventually fell riddled with bullets. There was little point in further bloodshed and his second-in-command, Captain De Peyster, surrendered 698 men. One hundred and fifty seven had been killed and 163 wounded, while the backwoodsmen lost only 28 killed and 60 wounded. All wars are unpleasant, civil wars particularly so. There was no magnanimity, and plenty of cruelty, in victory.

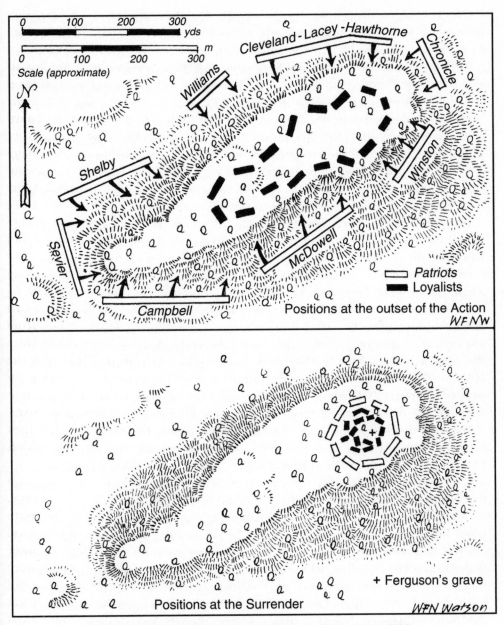

The Battle of King's Mountain, 7 October 1780

The disaster at King's Mountain tore Cornwallis's immediate plans for an invasion of North Carolina to shreds. Indeed, it would not be an exaggeration to say it was responsible for the loss to Britain of South Carolina. The death of such a good officer as Colonel Ferguson, and the surrender of so many men and arms were bitter blows, but the damage it did to loyalist morale and, conversely, the fillip it gave to that of the Patriots was incalculable.* A week later, Cornwallis withdrew his army to Winnsborough. The weather had broken, the roads were in an appalling condition, wagons sank in the mud, sickness was rampant and included the Commander-in-Chief, who handed over to Rawdon. To complete the despondent British picture, intelligence from Hillsborough reported that the dynamic and efficient General Greene had just taken over from the lethargic and incompetent General Gates.

Inevitably there were recriminations, although the existing evidence is unsatisfactory in determining their exact nature and when they were made. Undoubtedly Ferguson overstepped his orders in going well beyond the Catawba, for which Cornwallis cannot be blamed, and anyway, in a letter written seven years later, he says he had given him orders to retire. That is true, but it is uncertain whether Ferguson ever received the order, for he was writing to Cornwallis only a few days before King's Mountain, obviously unaware of any instructions to retire.[22] Clinton does not specifically censure Cornwallis for the débâcle, although in his Narrative he complains 'that during the whole of His Lordship's command he was certainly too apt to risk detachments without proper support.'[23] Tarleton is alleged to have blamed Cornwallis for not sending him to support Ferguson, but Cornwallis says that Tarleton pleaded sickness and would not go.[24] In the wider sense, Clinton felt sure that Cornwallis's precipitate withdrawal from Charlotte made the revolt in South Carolina more general,[25] and he was right. These may be fairly minor criticisms and censures, but they are significant in showing the undercurrent of mistrust that constantly flowed between the principals in this campaign.

Cornwallis was to remain at Winnsborough until the beginning of 1781, when he again marched north. He was in no position to do so earlier, for all around him the country was in turmoil. In most areas he was unable to offer protection to the Loyalists, whose numbers anyway were diminishing, there was much sickness in the army and, until reinforcements arrived in December, he could do little more than chase the guerrilla leaders and keep a watchful eye on the North Carolina frontier, from where reports were coming in of an American army assembling under General Smallwood.

In the south, Francis Marion, with the full support of a large majority of the local population, was fiercely oppressing what Loyalists remained and had succeeded in driving the British from Georgetown. Cornwallis sent Tarleton,

---

*Ferguson's body was treated shamefully, but when order was restored, the Patriots gave him a decent burial with, appropriately, a cairn to mark the site. It has remained undisturbed ever since, and can be seen to this day.

his most favoured trouble-shooter, to deal with him, but he was not long gone before information was received that Sumter, operating under orders from Hillsborough, was causing even more trouble in the area of Fishdam Ford, where he had worsted a small mixed force of Legion cavalry and mounted infantry under Major Wemyss. And so Tarleton was recalled to go north before he had met Marion. Accounts of the affair on 9 November differ, but it seems that Cornwallis must have been unwise to send Wemyss with only some 200 men against an experienced commander such as Sumter, who he must have known had at least double the amount of men. Even so, the result might have been different had not Wemyss's second-in-command, a young subaltern who was out of his depth in a difficult situation, abandoned the battle and his commander when the latter fell badly wounded.

Sumter now made for Ninety-Six, and, because he feared for its safety, Cornwallis urgently summoned Tarleton, who, marching at great speed, and gathering reinforcements as he went, gave chase. No sooner had Sumter realised who was after him than he aimed to shake off pursuit by crossing the broad, fast-flowing Tiger River well ahead of his opponents. This he might well have done had not Tarleton, after cutting to pieces a rearguard detachment on the banks of the Enoree, decided he must abandon his weary infantry and press on with just 170 cavalry and 80 mounted infantry.[26]

He caught up with Sumter, before he could cross the river, on the afternoon of 20 November. The Americans tooks up a strong position on an eminence called Blackstock's and, with double the strength of Tarleton's detached party, they had every advantage. Nevertheless, the fight was sternly contested, with Tarleton as always showing the offensive spirit, leading his cavalry knee to knee, sabres thrusting and hacking. But the enemy fire never faltered and the British were driven back in an orderly withdrawal. Sumter, although seriously wounded, remained in possession of the hill until the next morning when, certain that Tarleton would be reinforced, he crossed the river. Tarleton, who had lost a fifth of his men, had suffered his first major setback. He was to claim it as a victory, and in this he was supported by Cornwallis,[27] but it was nothing of the kind.

Cornwallis would not look back on the autumn of 1780 as one of the happiest periods of his life. He was not completely fit and his struggle to keep control of South Carolina was proving an almost impossible task. His main army was unbeatable, but improvised militia bands had everywhere proved desperately disappointing. Even Tarleton's professional pride had recently been humbled and, in the doing, Cornwallis's aide-de-camp, John Money, to whom he was devoted, had been killed. Then, only a few days later, on 1 December, Colonel Henry Rugely, a prominent Loyalist in whom Cornwallis had much confidence, surrendered to Colonel William Washington a strongly fortified blockhouse and its garrison without a fight, although he was in close contact with Rawdon's force at Camden. Cornwallis may have gained some satisfaction from the fact that Germain, encouraged by his success at Camden, was quite improperly beginning to communicate direct with him over Clinton's head. If so it was foolish, for it only inflated his ego and made for greater difficulties with his chief than those already existing, and could not but damage the campaign generally.

In the middle of October, Clinton had sent General Leslie to the James River at the head of some 2,000 men in response to Cornwallis's frequent requests for a diversion in the Chesapeake area. Clinton had instructed Leslie that he was to be under the immediate command of Cornwallis, who very soon 'invited' him to move to the Cape Fear River,[28] where he intended to meet him at Cross Creek. But by the end of November, the growing unrest in South Carolina, and the threat of an American army in the north, prompted him to bring Leslie's force closer. He therefore arranged for it to come to Charleston, where it arrived on 14 December.[29] Cornwallis, a week later, sent an extraordinary letter to Clinton in which he said 'The species of troops which compose the reinforcement are, exclusive of the Guards and the regiment of Bose, exceedingly bad.'[30] He had no grounds for saying this, *and he had not even seen the men*. Apparently he was anxious to cover himself, and his existing troops, against any future charge of inadequate performance relative to the increased numbers.

Clinton was rightly annoyed by this complaint, for it had not been easy for him to persuade Arbuthnot to keep back from the West Indies sufficient ships to convoy Leslie's men for what he hoped would be a profitable diversion up the James River.[31] It was bad enough that Cornwallis should 'invite' the force to leave the Chesapeake, but then to condemn most of it as 'exceedingly bad' produced a major explosion in the concatenation of Clinton's ceaseless complaints about Cornwallis. However, he still greatly desired the Chesapeake diversion. As we shall very shortly see, towards the end of 1780 a remarkable and quite unexpected turn of events was to present him with the right man for the task, so that he felt ready to accept the risk to New York of providing that man with the troops he needed.

The summer of 1780 in the North had been something of a stalemate, and most of what action there was had been at sea. Washington was very depressed at the continuing deplorable state of his army, and was only slightly encouraged by Admiral de Ternay's occupation of Rhode Island with 6,000 French troops under Comte de Rochambeau. Clinton did consider a counterstroke, but told Admiral Rodney he had not sufficient troops. This prompted a letter on 15 November from Rodney to Lord Sandwich in which he was extremely critical of both Clinton's and Arbuthnot's conduct of the war. He considered the evacuation of Rhode Island to be 'the most fatal measure that could possibly be adopted. It gave up the best and noblest harbour in America . . .' Had it not been evacuated, 'the French must have sheltered themselves in the Delaware or Chesapeake, where they could have been easily blockaded, which is not the case at Rhode Island, off of which it is too dangerous for squadrons to cruise in the spring, autumn or winter months . . .'[32] A little later, Rodney was to follow this letter up with a similar one to Germain in which he ends a paragraph of stern reprobation with 'his [Clinton's] affection for New York (in which island he has four different houses), induces him to retire to that place . . . and suffers himself to be cooped up by Washington with an inferior army without making any attempt to dislodge him.'[33]

Rodney, as we shall see in the next chapter, arrived in the West Indies as Commander-in-Chief in March 1780. He was a powerful figure in the Royal Navy and senior in rank to Arbuthnot (and, later, to Graves). It seems that he may have automatically assumed himself to be in overall naval command in the theatre of war. He acted in a very considerate and concilliatory manner towards Arbuthnot, but made his positiion (as he understood it) absolutely clear to him. Arbuthnot deeply resented interference and replied to Rodney in a most insubordinate manner, and continued to do so, thereby making any reasonable relationship between the two impossible. There is no record of the official position over Rodney's status but in the light of his consderable record and seniority, it is easy to see why he saw that position as he did. Nor does it come as any surprise that we should find him writing to the First Lord of the Admiralty and the Secretary of State for the Colonies as he did, although Clinton's conduct of the war on land was, in truth, no business of his. The letters seem to have been characteristic of that very distinguished but not, one judges, very lovable sailor.

No doubt his views on the value of Rhode Island were more realistic than Arbuthnot's had been, but to insinuate that Clinton was caitiff in not mounting a counterstroke on the island was unfair, for he genuinely did not have sufficient troops for an assault. Nor did he feel strong enough to carry out a project that was much closer to his heart, the taking of West Point, the newly completed stronghold up the Hudson. However, Benedict Arnold, who had virtually won Saratoga, had recently been transferred to command at West Point after a spell in Philadelphia recovering from his wound. There he had lived well above his means and had acquired a beautiful, but expensive, wife and considerable debts. For some time, he had had grandiose dreams of being the man who could end the war and had been in secret correspondence with Clinton through the latter's personal assistant, John André. Now, financial and family reasons gave a spur to his traitorous designs of surrendering West Point, and Clinton lent a willing ear. What followed has often been told, and no one concerned, with the honourable exception of André, came out of it with any credit. The plan misfired, André was executed, Arnold defected to the British and was made a brigadier.

Following the failure to win West Point by treachery, Rodney (according to his above-quoted letter) felt the need to open up the Hudson so strongly that he offered every assistance to take the place by force, but Clinton said it was too late in the season. In such an operation Clinton's appreciation was more likely to have been right, and anyway once the opportunity to obtain the place 'free of cost' had gone it was probably too late in the war to justify a large-scale attack on West Point and other posts which would need to be permanently occupied.

Meanwhile, Arnold on his arrival in New York immediately set about issuing proclamations and addresses to his former military and civilian compatriots, urging them to turn from the gullible persuasions of Congress and the French, but if anything his defection strengthened, especially in the army, the will to fight on. He then proposed an elaborate plan for the speedy reduction of resistance which London could not accept, and so Clinton was able to write to Cornwallis on 13 December announcing the imminent arrival of Arnold in the Chesapeake, adding 'I cannot suppose Your Lordship will wish to alter the destination of this corps without absolute necessity.'[34] Arnold sailed on 20 December with 1,400 provincial and militia troops for Portsmouth, from where he would operate up the James against American depots at Petersburg and Richmond. This it was hoped would take the pressure off Cornwallis by diverting a fair proportion of Greene's troops.

Thus the year 1780 comes to a close. In the North and the West Indies, British forces had done little more than hold their own, but in South Carolina there had been some pleasing successes and, although these had been overshadowed in the closing months, the year had ended on a note of cautious optimism.

Cornwallis, after a promising start, had been forced to abandon his march into North Carolina and was back at Winnsborough endeavouring to overcome the threat to his communications, which was largely due to Greene's intelligent policy. When that general took over from Gates at Charlotte he found his army in a desperate state. Scarcely 2,000 men were fit for duty, less than 1,000 of them Continentals, and they were short of arms, clothes, food and discipline.[35] The material deficiencies he was quickly remedying through the energy of Baron Steuben, in charge of the commissariat in Virginia. With that achieved, discipline would follow. Meanwhile, being clearly unable to meet Cornwallis in open battle, he had divided his army between the two principal guerrilla commanders, taking the larger force himself in support of Marion in the north-east, and sending General Morgan to the north-west to join Sumter. He thus created a threat to both Charleston and Ninety-Six, depending which way Cornwallis moved.

But however disturbing this was to Cornwallis in the short term, it did not prevent him from preparing for an early advance into North Carolina. Units were brought up to strength so far as possible, fresh horses were bought, the quartermaster's department generally, which had often displeased Cornwallis, was galvanised into some improvement, the defences of garrison towns were strengthened and their magazines increased. The arrival of Leslie's troops gave Cornwallis a comfortable superiority over Greene both in numbers and condition, making it difficult for the latter to detach troops to deal with Arnold. There was only one imponderable that could conceivably mar what appeared to be certain success in 1781 – would the expected Loyalist support be forthcoming?

*   *   *

Clinton opened the southern campaign with a strategical and tactical masterpiece in the capture of Charleston, but he and Germain had taken a calculated risk in launching it. This was the third time the main army had been split, and the fate of Carlton's and Burgoyne's armies did not augur too well for that of Cornwallis's, especially as in his case the distance between the two armies was much greater and the French were in the war. Even though the role of the northern army was defensive, and that of the southern offensive, there was still the problem of a divided command which created enormous difficulties in the overall strategy. But to the southern army a graver risk than strategical muddles was that the British navy would be unable to keep open at all times the vital supply line. D'Estaing had already shown that this could be a problem, and in due course de Grasse would ram the lesson home.

Matters were made no easier through the traducing and backbiting practised by the various principals. Clinton had found it difficult to work with Cornwallis (a difficulty which was reciprocated), or with Arbuthnot; Rodney was able to find fault with both of them, and Germain's behaviour exacerbated an already divided command. There was also the damaging effect of loyalist miscalculation.

Arbuthnot was old, and perhaps not very competent, although he had shown a touch of brilliance (attributed to Captain Elphinstone by Clinton) during the attack on Charleston. Clinton was always anxious for his removal and used his continued presence as one of the reasons for his constant appeals to be allowed to resign. In the end, Germain had relented sufficiently to promise Admiral Graves as his successor. Rodney, too, had reason to complain about Arbuthnot for withholding, incidentally at Clinton's request, ships needed for the blockade of Rhode Island, thereby spoiling Rodney's chance of defeating the French fleet. But the old man was not sitting down under that, and in a spirited letter to Germain he dismisses Rodney's general behaviour with 'all I can say is that Admiral Sir George Rodney displayed the most wanton unpresidented (sic) abuse of power that ever was exhibited.'[36]

Lord George Germain has been the principal butt of most historians for the mishandling of the war, and he was certainly responsible for much that went wrong in 1780. The unseemly wrangling between the various commanders was due, in large measure, to Germain's insistence on the wide dispersion of very limited resources; he undermined the authority of the Commander-in-Chief through issuing recommendations of a strategical and operational nature to Cornwallis, quite contrary to proper military usage; and he was in direct correspondence with other officers, notably Colonel Balfour, who sent him misleading information on probable Loyalist behaviour which created a dangerous optimism.

As the war expanded – first France, then Spain, and on 20 December 1780 Britain declared Holland a belligerent – Germain assumed an active interest in affairs beyond the Colonial Office, for he correctly considered that much of what other departments did came within his sphere. But this did not endear him to his colleagues, in particular to Lord Sandwich at the Admiralty, with whom he was seldom in agreement, to the detriment of naval warfare. And he summed up the chiefs in a note to Knox, 'How miserably we have been served in this war.'[37] Inevitably there was a strong feeling on both sides of the House that he should be removed, and protracted discussions between the King, Lord North and Lord Howe almost resulted in Germain succeeding North in the Wardenship of the Cinque Ports, but Germain insisted on a peerage and the King refused. And so he remained, and when all is said and done a suitable replacement would have been difficult to find.

The Opposition played the loyalist card with persistence, for they knew that the government was under pressure not to send more troops to America. This made ministers ever more extravagant with unwise assurances that their trust in sufficient loyalist support for future operations was not misplaced. There are those who assert that the failure of this support, and its consequences, have been greatly exaggerated. But the facts do not bear this out, for if it was right to say in 1775 that one third of the population supported the British and one third was neutral, as made clear in Chapter 2, those figures certainly did not read true by the end of 1780. Military reverses, unwise proclamations such as Clinton's of 3 June, and general disillusionment had swung some of the committed Loyalists and many of the uncommitted into the patriot camp. Successes in battle could never be consolidated without the willing co-operation of a sizeable number of Loyalists, and throughout most of the war this was significantly lacking.

The most serious disagreement among the chiefs was that between Clinton and Cornwallis, for it had a much greater effect on the war (especially that part of it fought in the south) than did the captious arguments that became a feature of Clinton's relationship with Arbuthnot. As we have seen, the two generals had never got on since the early days of the war. Indeed, it will be remembered that Cornwallis had at one point gone home in preference to serving under Clinton, and had only returned as his second-in-command because he thought there were prospects of succeeding him – although he would deny this.

There was no major disagreement over the sound strategy for the 1780 South Carolina campaign although, as matters developed, Clinton became irritated that his grand strategy for a sweep north from the Chesapeake became impossible, but he could scarcely blame Cornwallis for that. The trouble between the two was purely personal, and exacerbated by a distance of 1,000 miles. The time was not yet when a commander-in-chief could get into an aeroplane and within a matter of hours be on hand to iron out any misunderstandings with his field commander. When Clinton left Charleston in June he felt sure the subduing of the Carolinas was only a short time away, and he never properly understood what went wrong. He either did not know, or else he disregarded, Cornwallis's difficulties and his needs, and because he liked to have a scapegoat and had no great regard for his subordinate, he fell to criticising some of his actions without full knowledge of the facts.

Clinton, with his many and wide-ranging problems, lost sight of the fact that in South Carolina Cornwallis was the axle on which the wheels turned. For the time being, the province was his responsibility, and what mistakes were made he would acknowledge. Matters would have gone more smoothly had Clinton interfered a little less. In 1781, the road to North Carolina would be neither broad nor easy, and sadly these two generals were to tread it in an increasingly different climate of opinion.

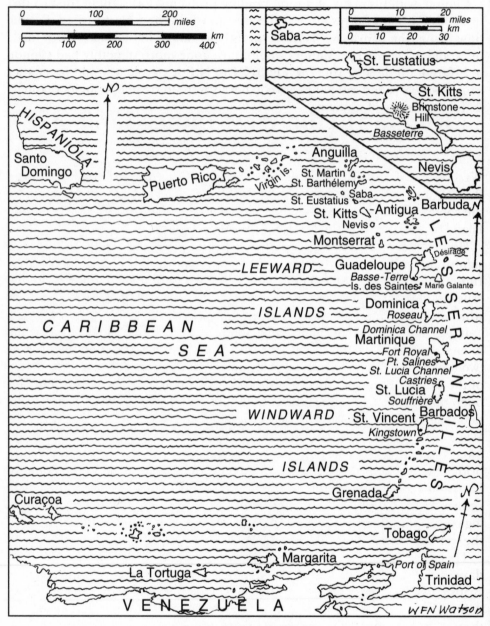

The Lesser Antilles

CHAPTER 11

Rodney in the West Indies 1780–82

WE SAW IN Chapter 10 how a powerful new naval figure, Admiral Sir George Rodney, had appeared on the scene in 1780 and had clashed with Clinton over the conduct of the war on land and, particularly, over the occupation of Rhode Island by a force under the French Admiral de Ternay.

However, Rodney's part in support of the mainland war was by no means as significant as his conduct of naval operations against the French in the West Indies, where he was Commander-in-Chief. These culminated in 1782 with the crushing defeat of their fleet in the Battle of the Saints in that year – a battle in which Admiral Hood too was to play a most important role.

It was that victory that finally persuaded the French that the time had come to end their protracted war with the British, following as it did on the heels of two other severe defeats, in European waters, in the previous year – the Battle of the Dogger Bank and the Second Battle of Ushant. Furthermore, their joint operations against Gibralter with the Spanish had proved a fiasco. So it may be said that the Treaty of Paris, signed in September 1783, marking the end of the War of American Independence and of the French conflict with the British, owed much to the restoration of British sea power. Although that naval story bridged a period of three years, it began chronologically at about the point we have now reached in the story of the war on the mainland of America. We shall return to the Carolinas Campaign of 1781 in the next chapter but it seems logical to follow the naval story to its conclusion at this point in the narrative of the war as a whole.

There had, of course, been spasmodic action in the Caribbean since the early days of the war; islands had changed hands, and the taking of St Lucia by the British in 1778 has been recounted in Chapter 9, but with the coming of Admiral Sir George Rodney and Lieutenant Général Comte de Guichen to these waters in March 1780, the fighting increased greatly in importance. Rodney, transferred from the Mediterranean, arrived at Barbados suffering severely from gout and gravel, on 17 March, just a few days before de Guichen brought his ships safely to Martinique. News of their progress out of Brest had been received at St Lucia via a fast frigate and although she had touched at Barbados, it was before Rodney had arrived and no message was left. It was therefore up to Admiral Hyde Parker to intercept the convoy, and Rodney was

169

annoyed that he had not done so. However, Hyde Parker somewhat redeemed himself when, on the very day after his arrival, de Guichen sailed for St Lucia in an attempt to retake that island, only to be thwarted by Hyde Parker's cleverly positioned squadron and the strong defensive works put up by the commander-in-chief, General Vaughan.

Meanwhile Rodney, still unwell, was determined to get his ships to sea and into action quickly. In de Guichen he had an opponent well worthy of him and the two fleets were fairly evenly matched numerically. However, the French ships were mostly in better condition being freshly out from home, whereas those English bottoms not yet coppered (see Chapter 14), and they were the majority, suffered severely from lack of proper maintenance through a paucity of dockyards in the Leeward Islands.

At first, Rodney could not get de Guichen to accept his challenge, but the latter was only waiting for the English to get clear of the Martinique coastline before he and the soldier-governor of Martinique, the Marquis de Bouillé, carried out their plan for a quick swoop on Barbados with 3,000 troops. Rodney soon had news of their sailing and put to sea with 20 ships of the line and five frigates. He met de Guichen's fleet of 22 ships in the early morning of 17 April and, for some time, the two fleets, sailing line ahead in opposite directions, crawled past each other with the British in a good position to windward. Manoeuvring went on until 11.50am when Rodney made the signal for every ship to bear down and steer for his opponent in the line.[1] His plan was to disable the enemy's centre and rear before the van, which was clear of the British line, could come to their assistance.

Unfortunately, British signalling was outdated and primitive, some signals for different actions being very similar. In consequence, Hyde Parker misunderstood Rodney's order and made off to attack the van, thus leaving Rodney's flagship, *Sandwich*, and the whole centre virtually unsupported. Their situation was rendered even more perilous when Admiral Rowley's rear division, having become slightly detached and been severely mauled by the French rear division, was unable to prevent the ships of that squadron from joining the attack on Rodney. Fighting went on until 4.15pm when, with the two fleets becoming vastly extended, the French broke off the engagement. The British ships had suffered so badly and were scattered over so many miles that any thought of pursuit was useless. Rodney transferred his flag to *Terrible*, for *Sandwich* had received such punishment that she had to be towed. The fleet remained at sea for some days, carrying out immediate repairs, before returning to Gros Islet Bay on 27 April. Thus ended indecisively the battle of Martinique, which Rodney felt should have been a complete victory had his captains, on some of whom he was to vent his displeasure, carried out his orders correctly.

De Guichen did not remain in Fort Royal long. On 6 May, news was brought to Rodney that the French were at sea again, sailing to windward of Martinique; their destination was unknown, but it was probably St Lucia. For four days, Rodney, with 19 ships, struggled against wind and current in the St Lucia Channel until 10 May, when his fleet was still only to the south of Point Salines on the southern tip of Martinique and de Guichen's fleet was sighted three leagues to windward. For the next few days, there was much manoeuvring with Rodney trying to bring de Guichen to battle and the latter refusing to be committed. Indeed, each afternoon he would bear down line abreast on the British line as if to attack and then, when just out of gun range, would haul his wind and retire. Both fleets had learnt tactical lessons from the battle of Martinique and Rodney had made sure that British signalling was improved. On 13 May, he transferred his flag to *Venus*, the centre frigate, whose duty it was to repeat signals.

By 15 May, Rodney had had enough of de Guichen's teasing and, through feigning retreat, lured him into a battle position. Soon both fleets were sailing line ahead on a parallel course. A sudden shift of the wind to south-east meant that the British were forced to turn to starboard to keep their sails full, but if the wind continued from that direction, it should ensure Rodney the windward gauge. But de Guichen saw the danger and went about, skilfully bringing his ships into line still to windward. At about 6pm, the wind backed to its original quarter and only about half the British line had 'turned the corner' to the north-east. Nevertheless, this manoeuvring had brought the lines close enough for Rodney's van to engage the French centre and rear as the fleets passed on

Naval Action of 15 May 1780

opposite tacks. For about an hour and a half the pounding continued, with both fleets suffering very considerable damage as the lines gradually drifted apart.

The two fleets remained at sea, with the French keeping a wary distance, until, on 19 May, there was another clash, again line ahead on opposite tacks. De Guichen once more had the weather-gauge and this time his ships crossed ahead of the British line, and there was an exchange of broadsides similar to that of the 15th as the two fleets passed each other, with the French rear suffering considerable damage from the British van. At last, both fleets had endured enough and, as if by mutual consent, the action was broken off. The fascinating tactical duel was over. The French had suffered the greater number of casualties in these actions, because their ships held a large number of soldiers for the attack on St Lucia. Rodney's casualties were 68 killed and 293 wounded.[2] A number of his ships were badly damaged, including the patched-up *Sandwich*, and *Cornwall* only just reached Gros Islet Bay where eventually she sank.

There was no serious naval engagement during the rest of the summer, although there was considerable anxiety in June when it appeared likely that de Guichen would be joined by a Spanish fleet. However, the Spaniards sailed north to Cuba. Rodney therefore had time to carry out a good deal of training

Naval Action of 19 May 1780

and to improve discipline in the fleet. This was very necessary even if it may have contributed to his unpopularity with his subordinates, all of whom recognised his ability although there were those who did not always give him the loyalty to which he was entitled. Shortly before the hurricane season put a stop to campaigning, Captain Walsingham arrived with badly needed reinforcements. In August, de Guichen sailed to France with the trade convoy and in the following month, Rodney, still not completely well and believing himself to be the senior naval commander in the region, went to America with half his fleet. There, as we saw in Chapter 10, he fell out with a truculant Arbuthnot, who resented what he saw as Rodney's interference.

On his return to the West Indies in December, he was appalled at the enormous damage done by one of those terrible huricanes that so often sweep across the islands and mainland of that region, leaving a trail of destruction and, usually, causing a considerable loss of life. Rodney found six of his ships had been lost, while others had been dismasted and there was a serious shortage of stores for repairing them. There had also been much damage on land to building and fortifications. However, there was good news that reinforcements under Admiral Hood were on their way. These comprised eight ships of the line, one of which was a 90-gun three decker, and all were copper bottomed and in good fighting condition.[3] The scene was therefore set for 1781, which was to be a momentous year in the war both at sea and on land.

In December 1780 war had been declared on the Dutch, and news of this was received by Rodney in the West Indies on 27 January 1781. He wasted no time in convoying troops in an attack on St Eustatius, the Dutch island where there was known to be a vast emporium in which was stored every kind of merchandise for the benefit of enterprising entrepreneurs, including traitorous British merchants trading with the enemy. The Dutch were taken completely by surprise, for the British sloop bringing news of the war had outsailed any ship that the governor might have expected, and he surrendered very quickly. The amount of booty collected was enormous, and its disposal was to involve considerable wrangling for many months to come, there being also 150 richly laden prizes to be taken in the roadstead. The smaller islands of St Martin and Saba were captured, but the opportunity to take other Dutch possessions (such as Curaçao) while they were still unprepared, was missed, according to Hood, because Rodney was too engrossed in collecting the spoils of St Eustatius. Hood's loyalty to his chief was not his strongest suit.

The next important event in these waters was the arrival of Lieutenant Général Comte de Grasse's fleet in early May (see also p215), thereby rendering the British navy numerically inferior at a critical time. The failure to prevent this was to form part of the Opposition's attack on Sandwich which will be described in Chapter 14; in his defence, Sandwich was to show that the fault lay not with the Admiralty, but with those who failed to carry out orders. He had written to Rodney on 21 March 1781 to warn him that a fleet, probably destined for North America, was being assembled in Brest and, as he said later in Parliament, there were too many commitments in home waters for it to be prevented from sailing. The responsibility therefore was clearly on Rodney's shoulders, but he was a very sick man at the time, depending largely on Hood. The trouble arose because he did not allow the very competent Hood to keep his ships where he had positioned them to windward of Martinique, with consequences described in Chapter 12. It is sometimes said that Rodney's orders to Hood to remove his ships to leeward of Martinique came from his desire to prevent the four battleships in Fort Royal from interfering with the convoy of valuable loot from St Eustatius that he was about to send to England. But there is no firm evidence of this, and it would not have justified putting the successful interception of de Grasse at risk.

De Grasse wasted no time in making difficulties for the British, whose various possessions were seriously threatened now that the French had naval superiority. Two days after his arrival, he took his whole fleet to St Lucia, together with 1,300 of de Bouillé's troops, which were landed at three points on the island. However, General St Leger's[4] open defiance and skilful dispositions, together with the hot reception given de Grasse's ships from the shore batteries when they anchored in Gros Islet Bay, determined the French once again that the island was too strong to be taken.* Meanwhile, Rodney had left St Eustatius and joined Hood's squadron off Antigua, from where they sailed to Barbados, which was considered the most vulnerable of the islands. Progress was slow, and it was not until 18 May that Rodney, who was without any news of the French, reached Bridgetown where he was relieved to find that all was well.

His ships were in very poor condition, but there was little to be done in the short time available, for on the 27th news was received that de Grasse, with a small squadron and 1,200 men, had made an unopposed landing on Tobago. The next day, Rodney sent Rear Admiral Drake with six ships of the line and three frigates carrying the 69th Regiment and two detached companies[5] with orders to return if he encountered the enemy in strength. Rodney has been criticised for not taking his whole fleet to succour Tobago, but he had four very good reasons for not doing so. The report was of only a small number of enemy ships; his own were in a very poor condition; Barbados, which was of more importance than Tobago, would have been left almost defenceless; and the fortifications on Tobago were considered strong enough to hold out for some time. And so they might have, had the inhabitants, who like many of these islanders were very half hearted in support of the war, not been threatened with the destruction of their plantations and, in consequence, persuaded the governor to surrender.

*This St Leger was Lieutenant-Governor of St Lucia in 1781. He could have been that Barry St Leger who was with Burgoyne in 1777. There is no *General* St Leger in the Army List of 1781; Colonel Anthony St Leger was three years senior to Barry, but his regiment (86th Foot) was not in America at that time, while Barry's (34th Foot) was.

Drake was back at Barbados by 2 June with information that de Grasse was at sea with his whole fleet. Rodney sailed for Tobago the next day but, on reaching the island, he learnt of the surrender. So he headed back for Barbados. On 5 June, he overtook de Grasse's fleet between Grenada and the Grenadines and, being to windward, was in a favourable position to bear down on the French and give battle that night. Hood, who did not have the responsibility of command, was adamant that it was a heaven-sent opportunity not to be missed, but Rodney would not be tempted; his fleet was inferior by three ships of the line and five frigates and, with those odds, he felt he should not risk either his ships or Barbados. Again, he was probably right. A few months later, Hood, then in command, was to make very much the same decision.

The early summer of 1781 was not a particularly happy time for the British in the Antilles or in the Gulf of Mexico, for Florida had fallen to the Spaniards on 9 May. It will be remembered that Brigadier John Campbell had command of a small number of troops in that area. In March he had been attacked by a Spanish amphibious force that built up over the next few weeks into a sizeable army. Campbell and his men fought most gallantly from their fortifications, but reduced to 650 and with their guns eventually out of action, they were forced to surrender.[6]

During the Caribbean hurricane season of 1781, the scene shifted to North American waters, and the operations that ended with Yorktown are recounted in Chapter 13. Rodney was in England during the autumn and early winter and Hood brought the fleet back to Barbados at the beginning of December. He was too late to stop the French from retaking St Eustatius and the other two Dutch islands, for although de Grasse was not back from America until the end of November, they still had ships enough to take advantage of Hood's absence. By the beginning of January 1782, Hood had 22 ships of the line fit for service, five of which had been lent by Digby commanding in America.[7] Furthermore, he had been informed that Rodney was on his way back with at least eight ships. De Grasse had 30 vessels fit for service and expected substantial reinforcements of troops and military stores to reach him soon. The French main plan for 1782 was the capture of Jamaica, but whilst they waited for the reinforcements to arrive, and taking advantage of their present numerical superiority, de Grasse and de Bouillé decided to attack Barbados with 6,000 troops.

The weather was particularly bad. During December, de Grasse, in two attempts, spent almost a fortnight trying to beat to windward through the St Lucia Channel before deciding to abandon Barbados and make for the easier, although less important, island of St Kitts. Hood remained at Barbados, uncertain of de Grasse's movements until mid-January when he received news of the French fleet from the governor of St Kitts. He immediately sailed north but, being in need of supplies and some repairs, he put into Antigua, which had the only good dockyard in the islands. Before sailing again on 23 January, he collected General Prescott and 500 troops. Reconnoitring frigates having reported to Hood that St Kitts was apparently invested, and de Grasse's fleet haphazardly at anchor in Frigate Bay, Hood hoped to surprise it at its anchorage, and probably would have done so had not one of his ships collided with a frigate in the night. Unwilling to lose the services of a 74-gun ship, he spent a day waiting for her to be repaired, by which time de Grasse had news of his approach and stood out to sea.

On the morning of the 25th, the French were a few miles to leeward. Hood, with the windward-gauge, might have attacked but de Grasse had 27 capital ships to his 22, and with the relief of St Kitts as his principal task Hood considered the risk unjustified. Moreover, there was a chance that he might seize the old French anchorage and, although the risk would be considerable and the manoeuvre demand the very highest degree of competence on the part of his captains, Hood decided to attempt it. While waiting at Antigua, he had explained to his two flag officers the tactics he would adopt in such circumstances and, in the action that followed, his trust in his subordinates proved well founded.

As soon as de Grasse realised Hood's design, he went about and bore down to the attack. However, it was two hours before he could come into action and meanwhile Hood's ships were nearing the anchorage. In the early afternoon, the battle began. Hood had outmanoeuvred his opponent and, under fire, his ships anchored in line of battle and brought their guns into action. It had been a magnificent display of seamanship, carried out with fire and spirit in the best tradition of the British navy. De Grasse made two attacks on the British line the following day, but his ships withdrew, having received from Hood's gunners more than they gave, and thereafter left the British alone.

Hood's squadron was to remain off St Kitts, a silent and impotent witness to the fall of the island his action had done so much to save. The island's garrison of just over 600 men was commanded by Brigadier General Fraser, a veteran of some years, but a stout fighter. On 11 January, he had withdrawn his troops before the French advance to a strongly defended position on Brimstone Hill, where he was joined by the governor of the island and some 300 militia. Here they were closely invested by the vastly superior French force which, although deprived of some of its heavy weapons through the loss of two ships, still had sufficient firepower to batter down Fraser's defences.

The islanders were not friendly (Fortescue,p408, says they captured some guns from the governor's militia and gave them to the French), and they saw to it that no information reached Hood as to what was happening on Brimstone Hill. It appears (again according to Fortescue) that Prescott was told by Fraser he could hold out without help, but it is not clear how the message arrived, or indeed what grounds there were for such optimism. The end was fairly inevitable: with their defences crumbling, their few guns out of action and the militia deserting, Fraser had no alternative but to capitulate to de Bouillé on 12 February 1782. The small garrison had fought most bravely for a month, losing 176 men of whom 38 had been killed.[8] With the fall of St Kitts, the dependent islands of Nevis and Montserrat were also lost.

When Hood eventually got definite news from the battle area, it was to tell of the surrender. It was now his wish to rejoin Rodney, whose return from England had been unexpectedly delayed owing to bad weather and a considerable degree of incompetence in the Plymouth dockyard. During the fighting on St Kitts, de Grasse had been reinforced and now had 33 ships of the line. It was therefore a matter of great importance for Hood's ships to cut their cables and slip out of Basseterre Roads without de Grasse, who was waiting ready with his superior fleet, being aware of their departure. This was done silently on the night of 13 February by careful synchronisation of watches, no lights, and a decoy boat left in harbour showing a light. De Grasse was suitably deceived and returned to Martinique, where he arrived on 26 February. Hood had met Rodney off Antigua the previous day and they sailed for St Lucia.

Rodney had brought with him 12 ships and five more were to arrive during the first half of March, giving the British 37 ships of the line and 11 frigates, the largest force they had had in these waters since the beginning of the war.[9] The two fleets would therefore be fairly evenly matched as the scene was set for the deciding naval contest of the Caribbean war. But first de Grasse had to receive a large convoy of much-needed stores. This convoy had put to sea in December 1781 under de Guichen, and been attacked by Admiral Kempenfelt with an inferior force. He had done it some damage, and the weather did the rest. Hence, on that occasion, only two transports had reached Martinique. It was now Rodney's task to ensure that the rest of the convoy did not get through intact on the second attempt. There seems to be little doubt that he made a mess of it.

Hood had received information that seemed to be reliable from two separate sources telling of a huge convoy which would leave Brest in February, destined for the East and West Indies. Sailing conditions were excellent at the time and, a Danish ship having made the trip in 29 days, it was to be expected that the convoy could be off Martinique by the middle of March. Rodney, who wanted all the time he could get for the repair of his ships, would not credit the French putting to sea before March, which was his first error. He then refused to countenance Hood's sensible suggestion that he split his fleet and thereby guard the two probable places of the convoy's entry, Désirade north of Marie Galante and the St Lucia Channel south of Martinique. The split would have given him sufficient ships to deal with the enemy at whichever point they were met, but Rodney felt it safer to keep the fleet together. He was convinced anyway that the French would come by their usual route to Fort Royal through the St Lucia Channel.

Hood put to sea on 16 March with 11 ships, for Rodney could not disregard the possibility of the French arriving earlier than he anticipated, but his orders to Hood restricted him to sail five to ten leagues directly to windward of Point Salines. On the 17th, Hood fell in with a frigate from whose captain he learnt that the French convoy might arrive at any moment, so Hood sent the frigate to St Lucia and Rodney sailed at once. For a week the fleet patrolled to windward of Dominica and Martinique, but saw nothing. Rodney relented on the 21st and ordered three of the line to take station off Marie Galante, but he cancelled the order the next day.

It was not until the 28th, that Rodney learnt that three ships of the line, three frigates and many transports carrying some 6,000 troops and a mass of military stores had reached Fort Royal safely on 20 March. They had made landfall at Désirade and, sailing south, had passed between Dominica and Martinique. Hood had been right, the French were not so foolish as to risk the heavily patrolled waters round the British base and he made quite certain that Lord Sandwich knew he had been right.[10] To us it might seem strange for the second-in-command of a fleet to vilipend his chief in a letter to their political master, but at this time it seems to have been almost obligatory.

De Grasse was now completely ready to sail with his large convoy of troops and military stores to Cap François, where the Spaniards were waiting with their fleet to join in the combined attack on Jamaica. Rodney was anticipating the event with satisfaction, for he felt confident he was capable of defeating the French at sea and ending the war in this theatre. He therefore welcomed information brought to him from a deserter on 5 April that troops were being embarked at Fort Royal and that the fleet was due to sail within a few days. De Grasse, on the other hand, was anxious to avoid battle, for his primary task was to get his convoy to Haiti intact. For this reason, presumably, he tried to slip out part of it quietly on the night of the 6th; but on sighting British vessels it returned. It was not until the 7th/8th that the whole fleet and its convoy sailed.

On the evening of Sunday the 7th, a frigate put into Gros Islet Bay with information for Rodney that part of the French fleet was under sail and that the rest would be by the next morning. Rodney found time to give the frigate's young commander a severe, but rather unfair, reprimand for quitting his station and was in no hurry to sail until the next day, which greatly surprised at least one of his captains, Lord Cornwallis's brother. On the evening of the 8th five ships of the line were detached to go forward to join the scouting frigates, and this small squadron caught up with part of the French fleet between the northern end of Dominica and the Iles des Saintes (which were to give the forthcoming battle its name) as they laboured to catch what land breeze there was. On the morning of the 9th, ships of Hood's division engaged 15 of de Vaudreuil's but later, as the breeze freshened, more came up and 20 ships on either side were in action. Three British ships suffered considerable damage, and one Frenchman had to be detached to Guadeloupe for repairs.[11]

The night was spent by the British in repairing their damaged ships, but the chase was resumed on the 10th and some progress was made. But de Grasse was quit of the slower transports, which had sailed westward under a small escort,[12] and he was further helped when Sir Charles Douglas, Rodney's chief-of-staff, made a signalling error that brought the van to a halt, and that night the fleet lay-to. However, Rodney was fortunate, for the 74-gun ship *Zélé* ran on board (the curious nautical expression for 'collided with'!) the 64-gun

Jason, and the latter had to retire to Basse Terre (Guadeloupe). When, next morning, Rodney saw *Zélé*, together with another damaged Frenchman, losing touch with their main fleet, he ordered his leading ships to close in on them, inducing de Grasse to bear down in support of his lame ducks. That was a mistake, for although Rodney recalled his ships, which were then in peril, de Grasse had lost much ground and would now find a general engagement virtually unavoidable. Matters were made worse for him when the unlucky (or perhaps mishandled) *Zélé* was in another collision that night, which put her completely out of action. This meant that for the battle which was shortly to begin, the British were the stronger, for Rodney had 36 ships of the line, while de Grasse, with these and earlier losses, could only manage 30.

Early on the morning of 12 April, Rodney decided that Admirals Drake and Hood should change positions, for Hood's division had borne the brunt of the fighting so far. After a complicated manoeuvre, well executed, Drake's division took the van and Hood's the rear. The two fleets now approached each other on a parallel course, the French, who were trying to form line as they sailed southward on the leeward side of Guadeloupe and the Iles des Saintes, managing to retain the weather-gauge. At about 7.30am, their van opened fire at extreme range, but the leading British ship, *Marlborough*, held her fire until the distance was down to 400 yards.[13] As the two lines began to draw level, Rodney sent urgent signals to close the range, and soon the two fleets were passing slowly at little more than a cable's length (200 yards). The French were receiving tremendous punishment from the British short-range carronades (short-barrelled cannon sometimes called 'smashers' from the havoc they wrought),[14] and de Grasse made vain efforts to disengage.

By now the smoke of battle confined visibility to not much more than the ship immediately opposite, where through the fog of powder smoke frantic gunners could be seen sponging out reeking barrels and ramming down new charges and shot to be sent pounding across the intervening water. But by about 9am, by which time the ships of Drake's van had mostly sailed clear of de Grasse's rear division, Douglas was able to discern through the gloom that seemingly all was not well with the French centre. A sudden shift of wind had caused some ships to lose their place in the line, and in the confusion it was plain to him that a gap had arisen between *Glorieux* and the next ship astern, *Diadème*, through which *Formidable* could lead others to seize the weather-gauge. Rodney took a little persuading to make this move but, once made, it proved the turning point of the battle.

As *Formidable* veered to starboard, she was followed by the next five ships, and all opened up with their port guns on *Diadème* and the three ships astern of her that had inadvertently closed up. The ship immediately ahead of *Formidable* in the original British line was the 90-gun *Duke*, and her captain shortly found himself sailing through another gap in the rapidly breaking French line. Farther astern Commodore Affleck, in *Bedford*, having passed a number of ships to starboard, found a third gap through which he led Hood's division.

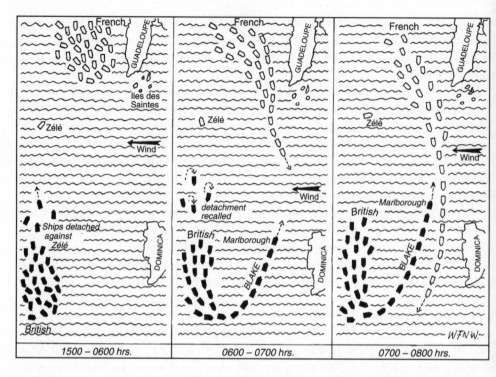

| 1500 – 0600 hrs. | 0600 – 0700 hrs. | 0700 – 0800 hrs. |

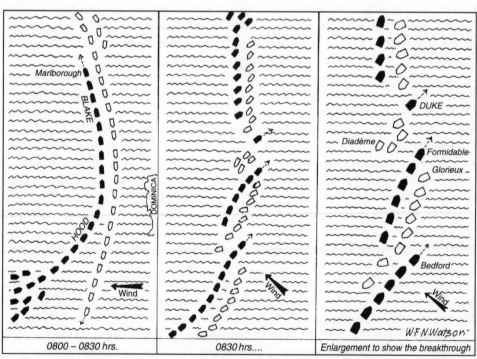

| 0800 – 0830 hrs. | 0830 hrs.... | Enlargement to show the breakthrough |

Naval Action off Les Iles des Saintes (The Battle of the Saints)
12 April, 1782

Firing continued until about midday, but for some time the French fleet had been in disarray and now, putting on every stitch of canvas, they were making off to leeward having had more than enough. The battle had been won, but the trophies had still to be collected, for only the totally crippled *Glorieux* had so far surrendered.

At a later date, Hood was to be most critical of Rodney's handling of the pursuit, for he was certain that had it been carried out more vigorously the squadrons of de Bougainville and de Vaudreuil would not have escaped to Curaçao and Cap François respectively, but Rodney was insistent that it should be controlled and that squadrons should not chase off independently. This may have been wise, for there was still fighting to be done that afternoon before *César*, *Hector* and *Ardent* (once a British ship, captured in the Channel three years earlier) were forced to lower their colours and, above all, before the flagship, *Ville de Paris*, pride of the French navy and a present from the city of Paris to Louis XV, was battered into surrender. De Grasse had lowered his flag and the flag of France at sunset, after his great ship had stood at bay with superb panache until her ammunition was almost expended and her decks were awash with the blood of the dead and dying. Her surrender accomplished, Rodney made a signal to break off the action, nor would he submit to Hood's entreaty the next morning that the pursuit be continued. British casualties in the battle had been surprisingly small. As always, figures vary. However, it seems that a few less than 250 were killed, and about 800 wounded. The French losses were very heavy, 400 being killed in the *Ville de Paris* alone, and in some other ships over 200.[15]

Undoubtedly the greatest gain from the victory in the battle of The Saints came from the capture of large quantities of armament and military stores destined for the attack on Jamaica. De Grasse had been captured, but the French were to rally under de Vaudreuil and could still muster some 26 ships of the line at Cap François together with a few Spanish ships, and the troops earmarked for the attack. However, the presence of Hood and Drake constantly cruising off the island well into May, and the loss of so much armament, ensured that the attack on Jamaica was abandoned. At home there was great rejoicing when news of the victory was received, and it was unfortunate for the new government that they had blundered into replacing Rodney at the height of his fame by the fairly mediocre Admiral Pigot. They were to make amends by offering Rodney a peerage, and anyway he was by now a very sick man; but that was not the reason for his recall, which was another example of unjustified interference in the services for political purposes.[16]

Pigot was not called upon to prove his talent, for after The Saints the French fleet showed no inclination to engage again, and their Spanish allies had always been half-hearted in their collaboration. But matters might have been very different had peace been delayed far into 1783, for largely through the efforts of Lafayette the French and Spanish governments had agreed to launch a huge combined armada to sail to the West Indies under the command of Comte d'Estaing. This fleet, which was eventually to consist of 66 ships of the line with transports for 24,000 men, was to assemble at Cadiz, and nine warships and 36 transports under Admiral de Vialis, with the Marquis de Lafayette on board the flagship, sailed from Brest for Cadiz on 4 December 1782 as part of the French contingent. This great enterprise fared even worse than other armadas, for after the signing of peace preliminaries it was broken up in Cadiz.[17]

* * *

The story of the West Indies in the closing years of the war mirrors that of Rodney and, to a lesser degree, Hood, for on the British side both these admirals dominated the scene. Rodney was an outstanding sailor, and if his performance in the West Indies occasionally fell short of the high standards expected of him, it can be attributed mainly to his being far from well for almost the whole time. Like everyone, he had his faults. He could be arrogant, greedy, short tempered and autocratic. He also had a dangerous dislike for delegation, which tended to dampen initiative in his subordinates, from whom he would never welcome suggestions or advice. This reluctance to delegate was the cause of considerable irritation to Hood, who would vent his feelings freely to colleagues and superiors alike in correspondence which more than hinted at his own greater tactical ability.

In spite of, or perhaps because of, these very human failings, Rodney was a fine fighting admiral, a thruster, a disciplinarian and a good trainer of men. His undoubted success in the West Indies might have been even greater had he had better health throughout, more seaworthy ships in the early stages and, most importantly, more time to inculcate his own brand of tactics and to update the signal system throughout the fleet. Certainly at the Battle of Martinique the result would probably have been different but for signal failure, and the old-fashioned ideas of his captains when ordered to bear down on the enemy line.

However, it has to be said that, in the period under review, few really bad mistakes were made. The one that had the most serious consequences on the course of the war was Rodney's failure to intercept de Grasse on his arrival in Caribbean waters in April 1781 and subsequently to allow him to sail to the Chesapeake, inadequately announced and pursued. This will be amply covered in the next chapter, but is worth emphasising. Hyde Parker had committed a similar error at the time of Rodney's arrival in 1780 but de Guichen, despite his undoubted skill, did not benefit greatly from this mistake during his short stay in Caribbean waters.

The navy was only indirectly involved in the loss of Tobago in May 1781 and of St Kitts in February 1782, for these were mainly land affairs. Their capture was largely due to that sempiternal problem of insufficient troops to fulfil the demands made upon them in a theatre of war notorious for the prevalence of debilitating diseases. Distribution of what troops there were may initially have been faulty, and Germain has had this added to the list of his misdemeanours, but there were many islands to be guarded and, in the last resort, their fate depended on command of the sea, which the British did not always have.

The inhabitants of the British West Indian islands were, at best, only lukewarm in their support of the war; their primary interest was commercial and some of them were not above trading with the enemy, especially when this was easy during Dutch neutrality, until Rodney put a stop to it. Their Assemblies were largely composed of planters whose knowledge of affairs seldom went beyond the bounds of their plantations, and they were very susceptible to the wooing and persuasion of the many American emissaries. As militiamen they were unreliable, for the safety of their properties was, perhaps understandably, more important to them than who occupied the island, but there were occasions – St Kitts for one – when they fought very bravely.

There had been no military commander-in-chief, as such, in the islands until the appointment of General Vaughan, who had arrived with two regiments at the beginning of 1780. General Grant had stayed for only a few months after his capture of St Lucia at the end of 1778 and he had had no time to arrange the proper co-ordination of the military effort, but with the coming of Vaughan this would be put right. It had been hoped in London that he could retake St Vincent and Grenada which had surrendered to the French the previous June, but he considered there were not sufficient troops, and decided that his first task should be to strengthen defences. For this purpose he had, in some of the islands, the local militia in support of the few regulars, and negro slaves were also enrolled and occasionally some of the Caribs.

In Tobago, where de Grasse had landed troops unopposed, the governor had 180 regulars, twice that number of militia and some armed slaves available,[18] and although he could retire to a strong position in the hills he was largely dependent on the militia and the planters. In the circumstances, faced with the importunate demands of the latter who were witnessing the destruction of their livelihoods, he had little choice but to surrender. As it happened had he held on for a few days the island might have been saved, for Rodney arrived with the fleet and supporting troops, but the governor was not to know this and, in the circumstances, it would be unfair to censure him.

St Kitts fell for very much the same reason, although in that island the population was actively hostile and gave assistance to the French. However, it is difficult to see how the outcome could have been different, for General Fraser's garrison was only about 600 regulars, and de Grasse had landed several thousand men. Undoubtedly Fraser had a very strong position on Brimstone Hill. Nevertheless, if it is true that he refused Prescott's offer of assistance, his reasons for doing so are not easily understood. Fortescue says of this 'Fraser was perfectly right. Brimstone Hill would have been quite safe but for the treachery of the inhabitants.'[19] But the odds against him were so great that it seems a crass mistake to have refused the offer of a relief force that could have taken the enemy in the rear, and their very presence might have had a beneficial effect on the froward population.

CHAPTER 12

The Carolinas: 1781

WHEN HE FIRST embarked on his South Carolina campaign, Clinton held the strategic initiative, and by the end of 1780 he still retained it. But there was much to be done in 1781 and the government looked for substantial successes that might herald the end of this tiresome war, for there was increasing disillusionment in Parliament and a dangerous restlessness among the London masses at the government's handling of both the war and the economy.

An election had been called at the end of 1780, for recent events, including better news from America, raised ministers' hopes of greater support. However, the results were not nearly as favourable as expected, for in a House of 552 members their notional absolute majority was only 26, weaker by five or six than in the previous parliament. These were days when parliamentary accountability to the electorate still lay in the future, and many members were nominees of influential patrons and had no constituency problems. In this election, 387 members were returned at the dictates of 93 peers, 104 commoners and a handful of ministers.[1] Where the elections were free and comparatively uncorrupted, pro-government support was less than anticipated, indicating a dislike of the American war; and although there were favourable results from some of the great trading centres, the government did not receive universal approval among the important industrial class of the large towns and cities from which Britain's ability to sustain a long war emanated. These men were increasingly alarmed at the state of the economy: the national debt had risen to over £250 million from a mere £127 million in 1775,[2] and the longer the war lasted the heavier became the tax burden.

However, the government had a mandate to continue its policies. Meanwhile, the Opposition lost no opportunity of attacking an administration that was becoming increasingly vulnerable. There was not only civilian discontent at public expenditure, rising prices and lower incomes, but there was incipient mutiny brewing when troops refused to go aboard ships sailing for America. To an unpopular government, military success was essential in a war that had recently been enlarged through a declaration against Holland, and much was hoped for from Clinton. This brought the spotlight on to Lord George Germain, who was the most hard pressed minister and the principal butt of the Opposition's rhetorical invective. He was a man in need of mending his ways,

The Carolinas and Georgia, 1780-81

but this he failed to do and continued to subordinate his undoubted intelligence and ability to a dangerous propensity for meddling in the structure of command.

There is no doubt that much of the estrangement that divided Clinton and Cornwallis throughout 1781 emanated from Germain's determination to treat the latter as the equal, if not the superior, of the former. Ever since Cornwallis's aide-de-camp had returned from England at the end of 1780, where no doubt Germain had intentionally or unintentionally made his true feelings clear, there had been a noticeable change in Cornwallis's behaviour towards his superior, verging at times on insubordination, and seriously interfering with Clinton's strategic planning.[3]

However, the worst of this lay in the future. At the turn of the year, Cornwallis and his army were ending a desperately unpleasant winter at Winnsborough. Food was very short, sickness was rampant and medical supplies almost nonexistent. Cornwallis had had more than two months to reflect on the events of 1780 and, while these had not been unsatisfactory, he knew that although his primary objective to safeguard Charleston had not so far been neglected, his approach to the subjection of North Carolina had not been along the lines that Clinton had advocated, and that province had still to be brought under control. But before that could be done, and South Carolina freed from the depredations of patriot guerrilla bands, Greene's army had to be defeated. And Greene had already stolen a march on him.

As already mentioned, there had been a great deal of hard work, and stirring up of the Quartermaster's department at Winnsborough throughout November and December, with the result that with more and better food, fresh horses and wagons, mostly impressed from a countryside that had of necessity been stripped fairly bare, the army was now in altogether better shape. Furthermore, by January, Leslie's troops were moving north through Camden, and Cornwallis reckoned that with them he had some 3,200 men (excluding garrison troops) which should be sufficient for him to defeat Greene. As it happened when that general had divided his force, and sent Brigadier General Morgan against Ninety-Six, he had given Cornwallis, with his larger army, the chance to defeat him in detail. Cornwallis planned to detach Tarleton to the west of the Broad River to deal with Morgan, while he himself marched up the Catawba with the rest of the army to rendezvous with Leslie near Turkey Creek, south-west of Charlotte. The combined force would then be at hand to block a retreating Morgan, and to swing against Greene whose men were on the Great Peedee in the area of Cheraw Hill.

According to his own account,[4] Tarleton was to take, in addition to his Legion's cavalry and infantry amounting to 550 soldiers, 200 men of the 7th and 71st Regiments, 50 dragoons of the 17th Regiment and two three-pounders, a total of about 1,000 men. This total, which is thought to be accurate in substance, although possibly not in detail, included three light companies. Their subsequent loss was a great blow to Cornwallis. He was confident that this force was quite sufficient to deal with Morgan who initially had around 500 infantrymen and 70 light dragoons under that redoubtable Patriot cavalry leader, Colonel William Washington.[5] Tarleton, who had begun his march a few days before Cornwallis, soon discovered that Morgan had abandoned his move against Ninety-Six, which place he decided could not be taken by his small force of mainly unreliable militia. Instead, he had wished to march against the less strongly held British outpost of Augusta. However, Greene ordered him to remain in the area of the Pacolet River to be ready to attack the British rear when they advanced into North Carolina. He had therefore headed north across the three parallel tributaries of the Broad – Enoree, Tiger and Pacolet. He got his men safely across all three, and left guards on the three fords of the Pacolet. In the event, Tarleton was delayed at the Enoree and Tiger, both of which had been swollen by recent storms. However, on 14 January, he crossed these rivers and arrived at the Pacolet on the 16th where his light companies found a crossing upstream some six miles from Morgan's camp at Thicketty Creek. From here, he sent a message to Cornwallis hoping he would march up the east bank of the Broad, for he expected to engage Morgan, who had made a hasty departure from his camp on learning that Tarleton was so close.

In the course of his march, Morgan had collected a good many local militia and he was now about equal to Tarleton in numbers, but he realised he was too hard pressed to cross the Broad and urgently sought a suitable place at which to give battle. This he found at Hannah's Cowpens, a fairly open, slightly rolling piece of ground where the back-country people pastured their cattle before sending them to market and where the backwoodsmen had rallied before King's Mountain. It was not ideal, but it had certain advantages, such as swamp and woodland on either side, which would prevent Tarleton from trying to turn his flanks. It also had the formidable Broad only five miles in rear which should have a steadying effect on the run-happy militia.

Morgan put 1,000 of these men in his first line with orders to stay put until they had fired at least two rounds; 150 yards behind them were 500 Maryland and Delaware Continentals, and in reserve were 120 of Washington's dragoons and 300 backwoodsmen.[6] Tarleton drew up his line within 250 yards of Morgan's militia, and placed his light infantry on the right with the Legion's infantry immediately to their left with one of the bronze three-pounder grasshoppers. On their left was the 7th Regiment with the other three-pounder, and slightly echeloned to the left and rear was the 71st Regiment. On each flank Tarleton placed dragoons, and 200 cavalry were in reserve.

It was 17 January and, as the sun rose on a bitterly cold morning, Tarleton's two guns sent over a few round-shot before he led the British front line himself against Colonel Andrew Pickens's militia. The British had been marching many miles across boggy and difficult country, and not surprisingly were very weary. Moreover, the regimental commanders, who had not even been consulted, were in some instances still posting their men, and quite unready to advance.

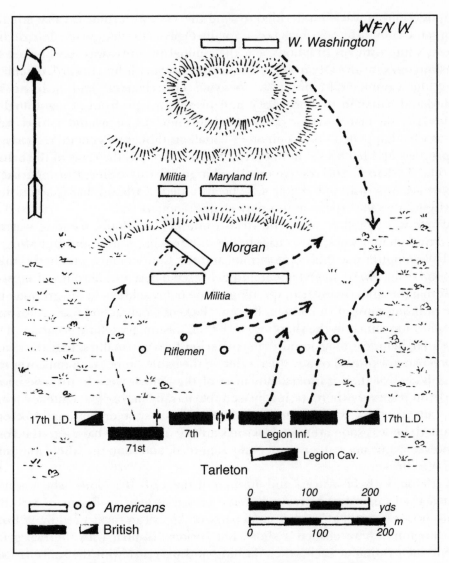

The Battle of Cowpens, 17 January 1781

But Tarleton was ever impetuous, and he firmly believed that his troops had only to show a modicum of resolution to sweep Morgan's ragged militiamen quickly into the Broad River. It must, therefore, have been something of a shock to him that the militia did not break on sight of the advancing British. Instead, making good use of cover, they did as ordered in firing two well-aimed rounds at the conspicuously attired officers and warrant officers before withdrawing.

The British line, now bereft of many of its leaders, was therefore a little unsteady when it took on the Continentals, who proved such a tough proposition that Tarleton decided to send the 71st, with some cavalry, to turn Morgan's right. He also ordered Captain Ogilvie to charge the left of the enemy's line with his dragoons, but their initial progress was soon halted by Washington's cavalry. Meanwhile, Lieutenant Colonel John Howard, commanding the Continentals, in order to avoid encirclement, had ordered his right-hand battalion to wheel back and present a new front at right angles, which is a risky manoeuvre in the heat of battle and can be misunderstood. And this is what happened now when other battalions thought a general retreat was in progress and fell back, but in perfect order. Here was the crisis of the battle, for had Tarleton's 200 reserve cavalry charged at this moment of indecision, the whole American line might have been routed. Tarleton, in his account of the fight,[7] says the order was given. If so, it was not obeyed.

Worse was to follow when the British infantry, no doubt scenting victory, tired though they were, gave chase to the withdrawing Americans. But Morgan and Washington saw their opportunity and the Continentals, who had retired in good order, were turned about and delivered a destructive volley of musket balls that tore into the British, spinning all too many soldiers to the ground and demoralising most of the others. By now, Pickens's militia had been reformed and they too joined in the shooting, with their customary telling accuracy. The 71st's Highlanders long stood firm, suffering heavy casualties, and the guns, which had not proved of any great value in the battle, were taken only over the dead bodies of their gunners. But most of the British were in total disarray. Tarleton was everywhere, trying by example to rally his badly frightened men. The air was thick with bullets and his horse was shot under him. Nothing daunted, he was soon atop Dr Jackson's animal and riding back into the fray. However, panic once afoot is not easily stemmed, and soon the whole line bent and then broke.

Tarleton, with 14 officers and 40 men of the 17th Dragoons, who were all that obeyed his orders to make one last effort, at least sufficient to save the guns, became entangled with a small party of American cavalry led by William Washington. There ensued a short, but furious, combat with the two great cavalrymen riding at each other slashing and parrying with the sabre before Tarleton, at last realising that all was lost, put a Parthian shot into Washington's horse and led his brave few from the field with some semblance of discipline based on professional pride.

Although the battle had lasted rather less than an hour, British losses were grievous, with some 800 men killed, wounded or captured, 100 horses and 35 wagons lost as also were the two guns and the Colours of the 7th Regiment. Morgan had only 12 men killed and 60 wounded. It was one of the very few tactical defeats suffered by the British Army in the war, and it had serious consequences for Cornwallis in that it virtually spiked his Carolinas campaign.

However, he was never to blame Tarleton, which was generous, for although that young commander cannot be said to have blundered, he certainly made mistakes. There was no particular reason to have hurried his troops into battle before they were ready, and while they were still terribly exhausted; many men of the 7th Regiment were raw recruits who, given a little more time, and properly supported by the guns, might have faced an entirely different situation with greater resolution. Nor can it be said that proper use was made of the cavalry, an arm in which the British had considerable superiority. Cowpens was not an engagement the flamboyant Tarleton would look back on with satisfaction, but it was only a temporary setback, and in no way blunted his brio or the value of his performance during the rest of the war.

At the time of Cowpens, Cornwallis had not, as Tarleton expected, drawn parallel to him in the area of King's Mountain where he might have cut off Morgan. Instead, he was at Turkey Creek, some 25 miles to the south-east. He had made the mistake of putting the Catawba between his force and Leslie's, which had caused delay. The news of Cowpens was a bitter blow, for with such heavy losses, in particular among his light troops, Cornwallis was well aware that an advance into North Carolina would be very hazardous. Nevertheless, as he wrote to Rawdon on 25 January,[8] he was determined to attempt it. Greene, on the other hand, was naturally elated by the events at Cowpens and, ordering his army north towards Salisbury, hastened to join Morgan, who had hustled his men across the Broad after the action so as not to be cut off.

In the course of the next two months, Cornwallis marched and countermarched his army many hundreds of miles across numerous rivers and creeks, most of which, as if through some supranatural Red Sea type phenomenon, became temporarily impassable directly the Americans had crossed them. His object was to get between Greene's army and Virginia before the Americans could receive the large reinforcements that were assembling in that province. He knew that Morgan had a good start of him and that Greene would be marching to join Morgan. So, to hasten his pursuit, Cornwallis took the drastic step of destroying his baggage and stores. Over a period of two days, while the army was assembled on the Little Catawba at Ramsour's Mills, officers and men stood by to witness the destruction of their personal belongings. Important luxuries such as casks of rum, all the stores, save ammunition, salt and medical supplies, and every wagon not required for the latter or for the wounded, perished in the flames of a huge bonfire. The army then marched for the Catawba where it found the Americans on the far side of a river recently swollen by torrential storms.

By 1 February, after two days of waiting, the river became passable, but most of the fords were guarded. During the enforced delay, Cornwallis had reconnoitred the area and decided on a demonstration of strength at Beattie's Ford, and on making the actual crossing at McCowan's private ford six miles lower down the river, which he felt might not be guarded. But in that he was wrong. Before the leading men of General O'Hara's brigade of Guards were halfway across the wide river, they were fired upon. There was no going back; the men struggled forward with great courage, lashed by bullets and unable to reply. The nearer to the far bank they floundered, the more intense became the firing. Cornwallis had his horse shot, but it reached the far bank before dying, and Generals O'Hara and Leslie also lost their mounts; some men were swept away, others were shot in midstream. However the leading troops eventually got across and the American force under General Davidson was engaged and dispersed, the general being killed.

The next objective was the Yadkin River. Morgan's lead was diminishing and Cornwallis had hopes of catching him before he crossed. However, when O'Hara's advance party reached the river, only the rearguard remained on the west bank. It was quickly dealt with and the baggage taken. The main body had got across in boats that Greene had previously organised to be available on wheeled platforms. Cornwallis, who had no such aquatic accessories, again found the river too swollen to cross. Clinton, writing with hindsight in his Narrative,[9] censured Cornwallis for not abandoning the pursuit at the Yadkin. His presence was badly needed in South Carolina, where British authority was everywhere being undermined by guerrilla bands. Although Cornwallis had dismantled the defences of Charleston, which seemed an unwise thing to do, he nevertheless was firmly of the opinion that he had left Rawdon quite sufficient troops to ensure the safety of the province and, at such a distance, he could be excused for seeing no reason to forego his great hopes of defeating Greene and bringing North Carolina back to its allegiance.

He now had to choose between waiting at Trading Ford for the waters to subside or marching his army upstream to a shallower crossing place. As the Yadkin flows through two large loops north of Trading Ford, it would necessitate a march of some 40 miles to the easier crossing, but Cornwallis – although perhaps not his men – considered it worth doing. The last obstacle for Greene, who had been joined by Morgan at Guilford Courthouse, before he reached the comparative safety of Virginia and reinforcements, was the River Dan. Throughout the whole campaign, Cornwallis had rightly complained that his intelligence service was lamentable, and now he had been informed that there was little chance at this time of year that Greene could get his troops across the Dan by the ferries, and that he could be brought to battle at the river. But when Cornwallis arrived at Boyd's Ferry, he learnt that the Americans and their baggage had passed safely across and that they had made very certain that no flatboats remained on either bank.

Cornwallis had had enough of river crossings and, at this point, he did give up the chase and marched back to Hillsborough. There he raised the royal standard and, in a proclamation, invited all loyal subjects to rally to the flag with arms and 10 days' rations. But he was not surprised when the response was small, for all knew that Greene's army was close by, and the appearance of his own was scarcely calculated to encourage would-be recruits to enrol. The troops had recently been subjected to long and gruelling marches over appallingly muddy tracks and were short of every comfort since the great bonfire. Even basic rations were not easily obtained and Commissary Stedman had to requisition and slaughter what few cattle and oxen could be found in the immediate area, an action, in itself, sufficient to alienate the local population.

After five days in Hillsborough, Cornwallis marched south-west on 25 February to camp in the area between Deep River and Allamance Creek which he thought would be a safer rendezvous for Loyalists. But this seeming retreat, at the same time as Greene's light troops under Colonel Lee reappeared in North Carolina, acted adversely on loyalist intentions. The fact that Cornwallis attacked Lee's men and drove them back across the Dan on 6 March made little or no difference to recruitment. Cornwallis had thoughts of marching farther south towards Cross Creek, where loyalist support should be better. However, on 11 March, Greene received his expected reinforcements and now, at the head of over 4,000 men, he felt strong enough to give battle. By the 14th, he had marched westward as far as Guilford Courthouse. (The Americans spell Guilford without the first d, and I have followed this except where the town is quoted in British army dispatches.) On receiving this information, Cornwallis immediately made preparations to meet him in the battle for which he had waited so long and marched so far to contrive.

The country round Guilford Courthouse was well wooded, but the trees had been cleared in the immediate vicinity, leaving open semi-cultivated land in two areas, each of about 500 yards square. These two clearings were divided by a defile where, for half a mile, the trees again closed in on the road. Greene had made good use of ground, but his position has often been criticised in that he had his lines too far apart and out of supporting distance, inviting defeat in detail. The North Carolina militia, positioned behind some interlocking rails, lined the north-eastern edge of the first clearing. Three hundred yards or so behind them, in the wood, the Virginia militia formed the second line. Between these two lines, which stretched across the road, Greene had positioned two bodies of cavalry and, on the road, two field guns. The third line, comprising the élite Maryland troops with the Delaware regiment and two six-pounders, was almost 600 yards behind in the angle formed by the Reedy Fork and New Garden-Guilford road. (The National Park Service Ranger, who has made a study of the battle, told the author that it was his opinion that the third line was at a distinct angle to the generally accepted position (see plan), and from walking the ground it would seem that he could well be right.) Greene has also been criticised for keeping his best troops too far back, and therefore unable to stiffen the militia. But whereas the distance between the lines may have been too great, there would not seem to be anything wrong with the disposition of the troops. Cornwallis's Intelligence, such as it was, reckoned that Greene had some 7,000 men defending this strong position, but in fact his total was probably only about 4,200. Even so, this was more than double Cornwallis's 1,900.[10]

The 15th of March 1781 was a day of bright sunshine, but there had been a lot of rain which made the going muddy and slippery for the British infantry, who had to advance to contact up a slight rise. As his troops crossed Little Horsepen Creek just before 1.30pm, Cornwallis could not discern the

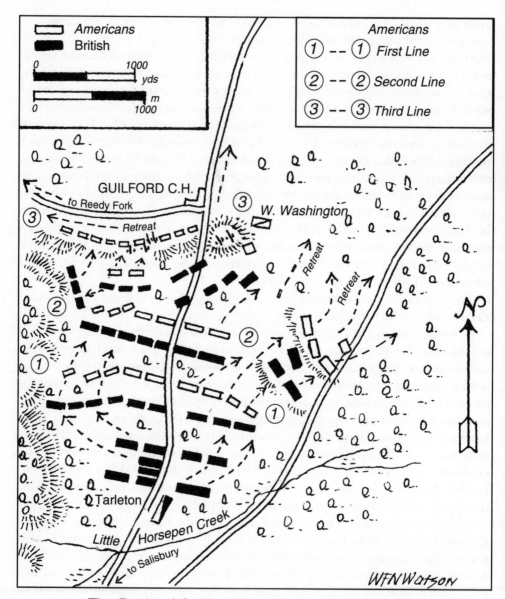

The Battle of Guilford Courthouse, 15 March 1781

American front line, but its position soon became obvious when the gunners opened up at about 600 yards. Cornwallis immediately deployed his leading battalions into line; General Leslie commanded the right with Bose's Hessians and the Highlanders, and the 1st battalion of Guards in support; on the left Colonel Webster had the 23rd, 33rd, and 2nd battalion of Guards. (There were two composite battalions of Foot Guards in America formed from all three regiments.) The guns, under Lieutenant John Macleod, with some jägers and light infantry remained on the road, and the guns were soon unlimbered and opening up; Tarleton's cavalry was in rear.

As at Cowpens, the first line militia had been ordered to fire two rounds before there could be any thought of falling back, and it seems that most of them obeyed their instructions, at least in part. But before the leading British soldiers came within range of the militia muskets, they had to endure a very unpleasant time advancing uphill over open ground under artillery fire and the longer ranged, more accurate rifles of Colonels Lynch's and Campbell's picked marksmen with whom Greene had stiffened his first line. Unflinchingly, these men strode up the rising ground, the air crackling with the whiplash sound of close-range bullets, until they reached the rail fence where they paused to deliver a withering volley into the North Carolina militia. Those who had still stood their ground found this volley too much and were soon gone, but the veteran riflemen remained steady on both ends of the line. Bose's regiment on the right, and the 33rd on the left, wheeled to their respective flanks to deal with these tiresome riflemen. The Guards, in the centre, slowly pushed back the American first line from which Green had by now withdrawn his guns and cavalry.

In breaking the first line, the leading British ranks had been shredded and exhaustion was beginning to lay a dead weight on all. Now it was to be a confused woodland fight and, with the possible exception of the jägers and light companies, the American soldiers had the edge over the British in this sort of country. Cavalry and guns were of little use, and even the bayonet was at a disadvantage, the fighting being mainly hit and run, dodge and dive work. Cornwallis had to bring up his reserves. For apart from casualties, Bose's regiment and the 1st battalion of Guards were still fighting (and would continue to do so for the rest of the battle) away on the British right against Campbell's and Lee's men, who had proved so stubborn on this flank during the attack on the first line.

Wherever the fighting was fiercest, and his presence most needed, Cornwallis was sure to be there, his coolness of judgement never impaired by the confused nature of the battle. Stevens's Virginians on the American left and Lawson's on their right fought splendidly. But gradually, Lawson's brigade was rolled back on to the road and from there the troops retired in some disorder onto the third line. But Stevens's men, despite seeing their gallant commander carried from the field, fought on, giving as good as they got, until, being forced into the open, they were dispersed by Tarleton's cavalry.

And so, at last, the American third line was reached. The battle remained in equipoise but those weary British soldiers who still survived now faced their most formidable task, for Greene had kept his best troops until last, the renowned 1st Marylanders and Delaware men. Webster opened the British attack on the left, but the Marylanders drove his men back in disorder and with heavy loss. Here Greene may have missed a chance of victory in not following up this temporary rout, but who can be certain whether a chance exists in

fighting of this kind? Anyway, Webster's men soon rallied and, aided by
Macleod's guns and the 2nd battalion of Guards, resumed the offensive.
General Leslie also brought up the 23rd and Highlanders in support. The
battle now became a moiling mêlée of close-quarter fighting with every man
striving to kill before being killed.

The 2nd Marylanders, a young regiment recently recruited, found this
introduction to battle unsatisfactorily warm and departed the field, the Guards
in pursuit, capturing two six-pounders. However, on being counterattacked,
they were forced to abandon them. At this stage, Washington's cavalry, from
their commanding position on the left of the American line, swooped down on
the rear of the Guards and the British situation had become desperate.
Cornwallis was quickly on the spot. To stop the cavalry from tearing the entrails
out of his whole line, he order Macleod to fire grape into the confused medley.
This broke up Washington's dangerous onslaught, but also caused grievous
casualties among the Guards.

The terrible battle was now nearly over. Cornwallis availed himself of a pause
in the fighting to reform his line. The indomitable Webster, who had already
received a ball in the knee and was now to be more seriously wounded, led the
33rd once more into the attack, while O'Hara, who had also been wounded,
reformed the 2nd Guards battalion and brought it up alongside the 33rd. On
the right, the 1st Guards, Bose's regiment and Tarleton's cavalry were at last
driving the Virginia militia off the field. Greene saw the battle was lost and gave
the order to retire, abandoning his guns because he had no horses left to draw
them. He marched 15 miles through the night to his camp at Troublesome
Creek, leaving many men dead and even more wounded, whose pitiful chorus
of screams and groans could not be expunged as night and rain cast a sodden
cloak over the carnage of the field.

Cornwallis could claim the victory, for his army occupied the ground, but it
was a victory to mourn over rather than to rejoice. His casualties had been
devastating. Including the gallant Webster, who was to die during the
withdrawal south, the British had lost five officers, 13 sergeants and 75 other
ranks killed; 24 officers, including two brigadier generals, and 389 other ranks
had been wounded – many of them seriously – and 26 men were missing.[11]
This total of 532 was more than a quarter of the men engaged. Cornwallis's
army was ruined and the Carolinas all but lost; however, the battle of Guilford
Courthouse stands high in the annals of the British Army. The heroic troops,
superbly led but half starved and very weary after a long march, had defeated
an army of more than twice their numbers that was in better condition, better
armed and strongly posted.

It is legitimate to ask whether Cornwallis should have entered upon this
battle, or whether he would have been better advised to have fallen back on
Camden to join with Lord Rawdon, for Clinton's most explicit orders to him
were to safeguard South Carolina. When Greene recrossed the Dan, his army
was reported (wrongly as it happened) to number 7,000 men, while Cornwallis

had barely 2,000 fit for duty, and only partly fit at that, for they had been on short rations and hard marching for a considerable time. The odds were heavily stacked against him and, in the end, it was only the fighting quality of his troops that saved the army from more serious destruction. On the other hand, he had set his heart on bringing North Carolina back into the royal fold. To do this, he had to achieve two things. He had to win the hearts and minds of the majority of the colony's people, and he had to defeat Greene, but sadly, as Cornwallis knew, as he would later admit in a letter written to Germain from Wilmington,[12] the first had by now become impossible without the second. With hindsight, it is possible to see that at Guilford Courthouse the British lost South Carolina although, judging from his subsequent action, Cornwallis did not think so. However, his wish to accept battle is entirely understandable, for he had marched so far and suffered so much for this chance to 'put it to the touch' and win all. It is easy to say he was wrong – and he probably was.

There could be no question of the victors pursuing the vanquished. Indeed, it was to be a case of the vanquished pursuing the victors. Cornwallis's much diminished army was in a dreadful state. The men had not eaten for over 24 hours, and then only a quarter of a pound of flour and some stringy beef, for Greene's cavalry saw that no local provisions were available and nothing could come up the rivers. Inevitably, there was sickness born of undernourishment and battle fatigue, and there were over 400 wounded to be cared for. There was no alternative for Cornwallis but to march south, although there was a choice of places to go to. So, having spent a day sorting out those wounded too ill to be moved (about 70 of them), loading the others on to carts, and making humanitarian arrangements with Greene for the wounded of both sides, the army departed on 18 March for Cross Creek – with the captured cannon and those prisoners not on parole.

Before he left, in a totally hopeless gesture, Cornwallis issued a final proclamation. In this he informed all would-be Loyalists that 'by the blessing of Almighty God his Majesty's arms have been crowned with signal success, by the compleat [sic] victory obtained over the Rebel forces . . .', and he called upon them 'to stand forth and take an active part in restoring good government'.[13] As it must have been well known that Greene's army was in the neighbourhood and quite intact, whereas what was left of Cornwallis's was even then preparing to depart many miles to the south, this proclamation was a foolish mistake, productive only of mirth and ridicule.

On arriving at Cross Creek, another disappointment awaited Cornwallis, who had been hopeful that ample provisions would be available, and that the large and loyal Scottish colony there would swell his ranks. In the event, there was no food to be had and precious little loyalty. As he was to explain to Germain in subsequent dispatches, there were insufficient provisions within 20 miles; the Cape Fear River was not navigable to his nearest supply base, Wilmington, and 'the inhabitants on each side almost universally hostile'.[14] Moreover Greene, with a well provisioned army, was coming after him. There was nothing for it but to plod on to Wilmington. The condition of his small army had become quite pitiful, many of the men marched barefooted and all of them looked like grey ghosts, their faces reflecting despair. Many of the wounded (including Colonel Webster), bumped about in carts and with no proper medical attention, died painfully, while the few who still had some strength ransacked what properties they passed on the right bank of the river in the forlorn hope of finding food. On 7 April, after 18 ghastly days on the road, the army stumbled into Wilmington.

The difficulty of communication between the Commander-in-Chief in New York and his virtually independent army commander in the Carolinas had always been pronounced, but had become far worse when Cornwallis marched his army into North Carolina several hundred miles from his sea base. Consequently, Clinton had no control, and was quite unaware of what had been happening until he received Cornwallis's dispatch sent from Wilmington on 10 April. This began by informing Clinton that 'our military operations were uniformly successful; and the Victory at Guildford, altho' one of the bloodiest in this War, was very compleat'. There was nothing in this dispatch to indicate that the Carolinas were as good as lost, although a warning bell might have been sounded by rather ominous sentences such as 'Until Virginia is in a manner subdued, our hold of the Carolinas must be difficult, if not precarious', and again 'North Carolina is of all the Provinces in America the most difficult to attack'.[15] But no misgivings appear to have been aroused, and Clinton was at first – as they were in London – exhilarated by his subordinate's achievements.

However, it was not long before full information was available and opinions changed very considerably in both New York and London. Fox, in the course of two great speeches in the House on 12 June, did not spare Cornwallis. 'The truth was, the victory of Guildford, as it was called, drew after it all the consequences of something very nearly allied to a decisive defeat. Lord Cornwallis did not fly from the enemy; but indisputable facts bore him [Fox] out in affirming, that if Lord Cornwallis had been vanquished, instead of being the temporary victor, his operations, or rather movements, could not have borne a more unfortunate aspect.' Cornwallis had, Fox declared, 'relinquished all the advantages he had gained with so much difficulty'.[16]

Clinton was more specific in his Narrative and, although this was written later, it is fair to assume that the manuscript reflected his thoughts on learning the full situation. The main burden of his complaint was Cornwallis's seeming disregard for the safety of South Carolina. He admitted his approval of the march into North Carolina, so long as it was underpinned by a good Loyalist response; he was adamant, however, that when this became nothing but an illusive mirage, as it had by the time the Yadkin was reached, Cornwallis should have pulled back. Finally, on reaching Cross Creek and finding it bereft of supplies and Loyalists, Clinton suggests that the army should have taken the better, and scarcely longer, road to Camden where, with Rawdon, he had every chance of complying with his orders to safeguard the province. It is difficult to disagree with this criticism and not to conclude that Cornwallis committed some grave errors in the latter part of the campaign.

He did not stay long at Wilmington. Instead, he compounded his disobedience by marching north for Virginia on his own authority. Three days earlier, he had learnt that Greene was posing a direct threat to Rawdon, whom he had put in charge of the 7,000 troops still in Carolina and Georgia. It would have been comparatively easy for him to have returned either by sea or overland, but he had become convinced that the war could be won in Virginia and that his presence was needed there. In the letter to Clinton, already quoted, he had even suggested that the Chesapeake might become 'the Seat of War, even (if necessary) at the expense of abandoning New York'. To which Clinton was to add a marginal note, 'It certainly would have been the speediest way of finishing the war – for the whole army could probably have been annihilated in one campaign commencing in July'.[17] And so, on 25 April, Cornwallis left Rawdon to his fate and marched north at the head of some 1,500 men and four guns. He sent an explanatory letter to Clinton ahead of him[18] in which he said that, unless he moved quickly, Greene might cut him off from General Phillips, recently arrived at the Chesapeake. He also sent another letter to that general,[19] who was a personal friend.

Meanwhile, Greene, who had given up his pursuit of Cornwallis at Ramsay's Mill on the Deep River, sent Lee to join with Marion in an attack on Fort Watson, an important post on Rawdon's line of communications some 50 miles north-east of Charleston. This they accomplished by resorting to the medieval siege practice of erecting a large wooden tower from which to pour not boiling oil, but well-aimed rife fire, on the 100 or so men that manned the stockade. They then rejoined Greene, who, by 19 April, was before Camden, which had been strongly fortified after the 1780 battle with a stockade wall and a series of six redoubts around the perimeter. Greene decided it was too well defended for his men to make a successful assault and withdrew a short distance north to take up a good position on Hobkirk's Hill. Here, on 25 April, Rawdon, having received information that some of Greene's militia were away on escort duty, decided to attack him while numbers were more equal. Even so, the British, with only about 800 men, were at a numerical disadvantage, without any guns and pitted against some very good Continental troops. However, Rawdon gained a measure of surprise through taking a circuitous route to come in on Greene's flank.

The tactics of this short battle are interesting. Immediately Greene had turned his line to meet Rawdon's flank attack, he opened up his two centre battalions to reveal his guns, which at once discharged grape into the British leading ranks and caused them to fall back. Greene then ordered his cavalry to wheel round and come in on Rawdon's rear while his two flank battalions attacked the British flanks, and his centre battalions, having closed ranks again, advanced against the British centre with the bayonet. But Rawdon was not to be caught. He had good troops posted on his flanks and he, in turn, opened up his

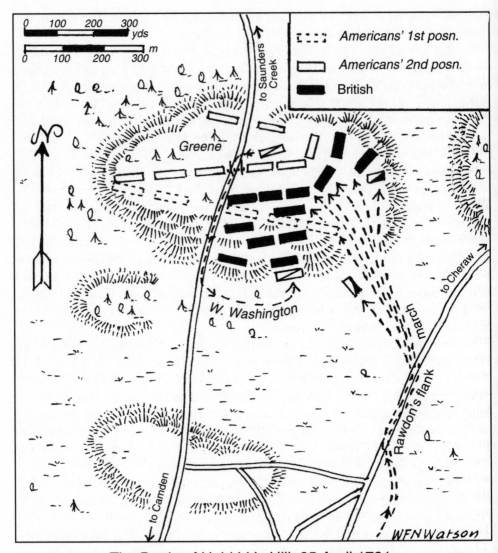

The Battle of Hobkirk's Hill, 25 April 1781

centre to let the Americans come through. He then closed the trap. Inside this cauldron, the fighting was very fierce, with Rawdon's infantry lapped about by Washington's cavalry in the never ceasing process of death and mutilation.

It was the loss of many of their senior officers that eventually broke the Marylanders' battle spirit. Once one battalion withdrew, others soon followed. The end came quite suddenly, with Greene and Washington personally manhandling the guns to safety. Rawdon had lost a little over 200 men. Nevertheless, in this, his first senior command in battle, he had displayed considerable tactical skill – the harbinger of greater things to come. Clinton was to say that Hobkirk's Hill 'was perhaps the most important victory of the whole war, for defeat would have occasioned the loss of Charleston (in the open state of the works of that capital), the Carolinas and Georgia'.[20] It is tempting to think that this hyperbole in his Narrative was prompted by the desire to detract any credit from the overall performance of Cornwallis in Carolina; nevertheless, Rawdon's was a fine effort by a young commander.

In fact, Hobkirk's Hill, did no more than slightly delay the loss of Georgia and the Carolinas, whose fate had been sealed at Guilford Courthouse. It was not long after the battle that Greene and his guerrilla chiefs, operating on interior lines, initiated the departure of the British by running rings round their comparatively isolated posts, for whose protection Rawdon had insufficient troops. Camden itself had to be abandoned on 9 May before Greene closed the net around it, and Rawdon fell back to Monck's Corner, where he was well placed to receive supplies from Charleston. At the same time, the guerrilla Colonels, Marion and Sumter, and Pickens's militia, were busy capturing a number of lightly held forts on the rivers Congaree, Wateree and Santee and, early in June, Colonel Lee laid siege to Fort Cornwallis at Augusta on the Savannah River. This post was resolutely defended by Colonel Browne and his Loyalists but, following its success at Fort Watson, the wooden tower idea was repeated and all Browne's gunners had soon been picked off and he was forced to surrender. And so, only a very few weeks after Greene had appeared before Camden, the British had nothing left in Carolina save Charleston and Ninety-Six.

At Ninety-Six, Colonel Cruger had a garrison of some 550 Loyalist, provincial and militia soldiers, 200 of whom, together with three 3-pounders, were posted in the main star-shaped fort to guard the east side of the village, with the remainder spread between the jail and a stockade fort to the west of the village. These two outposts were connected to the Star Fort by covered ways. The surrounding ground was open, and the road to Charleston, called Island Ford Road, ran to the west of the fort and straight through the village. The garrison was isolated in the midst of enemy territory and, before its investment, Rawdon had seen the need to evacuate the position. Messengers had been sent to order Cruger to withdraw, but so great was the Patriot presence in the area that none of these messengers got through. Consequently, Cruger was totally unaware of the general situation in South Carolina.

Greene himself took charge of the attack, having with him about 1,000 regulars, some militia and four six-pounders. He arrived before the position on 21 May to find that a great deal of work had been done on the defences, and efforts in that direction had recently been redoubled. He began by bombarding the Star Fort and, during the night of the 21st, he erected two works within 70 yards of it. However, he was quickly made aware that he was not dealing with raw troops when a platform was mounted to take the garrison's three cannon and, under cover of a bombardment, a sally party of 30 entered the American

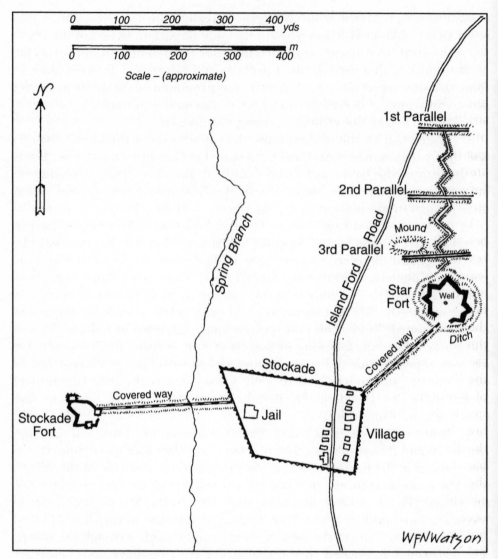

The Siege of Ninety-Six, 22 May – 19 June 1781

works and killed every man in them. Greene then resorted to the regular pattern of siege warfare and the first parallel was begun at the safest distance of 300 yards. By 3 June, the second parallel had been completed and batteries brought up. A commanding mound made of gabions (woven baskets of brushwood filled with earth) had also been constructed which enabled the Americans to fire down upon the garrison.

At this juncture, Cruger was summoned to surrender, but every man, fearful of the consequences, unhesitatingly opted to fight on. Thereafter the bombardment intensified, the third parallel was begun, saps were thrown forward for mining the Star Fort and, with the besiegers reinforced by Colonel Lee's troops from Augusta, Greene was able to enlarge his operations. As the siege progressed, the garrison inevitably suffered shortages of food, ammunition and, above all, water. A well was sunk in the Star Fort but found to be dry and, because the only water supply for the two forts and village was the local stream, now in the American lines, naked negroes, blending with the darkness of the night, went on perilous missions to fetch sufficient supplies to ensure the garrison's survival. It seems incredible that a village of 12 houses, a courthouse and a jail did not have a well from which water could have been brought to the forts by covered way. But historical records and a visit to the site confirm the stream as the only source.

Despite the efforts of these brave men, by 18 June the garrison's plight was becoming fairly desperate, until a messenger from Rawdon got through who revived spirits with the news of a relief force. Rawdon had only been able to leave Charleston by the fortuitous arrival of three regiments from Ireland originally destined for New York, and he had begun his march on 7 June in heat so appalling that 50 men were to die on the way. (It was asking a lot of these reinforcements to land them in Carolina in the torrid month of June. On this occasion it was not bad planning, for they were destined for New York, but too often little thought was given to time and climate when sending troops to the south, and particularly to the West Indies.) Greene, too, had become aware of Rawdon's march, and set in motion one last effort to take the fortress by storm. By 10 June, his third parallel had been completed, mines had been laid, breaching batteries brought up and zigzags sapped forward so as to put men in the ditch ready for the assault. These preparations could have brought success, had it not been for the élan and courage of two parties, each of 30 men, who on the night of 18 June issued out from a sallyport in rear of the Star and swooping into the trench from two directions bayoneted two-thirds of its occupants. This, and the close approach of Rawdon, decided Greene that it was now time to raise the siege and, on 20 June, he marched his army across the Saluda River. The relief force arrived the next day, and at once went in pursuit as far as the river Enoree, but Greene was marching too fast and Rawdon's men were suffering greatly from the heat. So they returned to Ninety-Six to cover the evacuation of the gallant garrison.[21]

Shortly after Rawdon had withdrawn his army to Orangeburg, and brought the Loyalists of Ninety-Six district within the protection of the British pale, he returned to England on sick leave. He would be greatly missed, and it was sad that further opportunities could not be given him to show what was undoubtedly a burgeoning military talent in a theatre of war where such ever growing quality and performance were badly needed on the British side. The command in South Carolina now devolved on Lieutenant Colonel Alexander Stewart of the Buffs, who had arrived with the recent reinforcements.

After his hasty withdrawal from Ninety-Six, Greene spent the hot months in the high Hills of Santee, while Stewart remained in the area of the Congaree River. Meanwhile, on being reinforced by some North Carolina militia, Greene moved in very easy stages towards the British army, which Stewart pulled back to Eutaw Springs to be at hand to receive a convoy of provisions. On 7 September, Greene was joined by Marion's brigade, and early on the 8th he marched for Eutaw. Stewart's intelligence was even worse than Cornwallis's had been; Greene's approach took him completely by surprise, so much so that he had a party of about 100 unarmed men[22] with just a light cavalry escort foraging for sweet potatoes some two miles in advance of the camp.* Moreover, when, at 6am on the 8th, two American deserters brought news of Greene's march, they were not believed. It was not until three or four members of the foraging party (the sole survivors) galloped into camp in disorderly fashion, that Stewart realised that the Americans had spoken truly.

There was little time for him to draw up his line of battle, but the ground more or less dictated his dispositions. The camp was in a fair sized woodland clearing at the back of which was a house and garden. Stewart relied on a single line placed just beyond the clearing, with his right flank resting on Eutaw Creek, where Major Marjoribanks and 300 men were positioned slightly ahead of the line and at an angle to it. The left flank reached across the road down which Greene was advancing; two guns occupied the road and the cavalry were slightly in rear on the left of the line. An officer (Major Sheridan) had been detailed to occupy the house should this become necessary. Greene advanced in two lines with his militia and two guns in the first, and three regiments of Continentals in the second. His cavalry under Washington, and the Delaware regiment brought up the rear. In all he had some 2,000 men, which was more than twice Stewart's force.

The action began with a lively cannonade that put one of the British guns and two of the Americans' out of action. Greene's militia, advancing with great steadiness, then closed with the British and it was some while before those troops gave ground under the withering fire. Stewart's left battalions then rashly surged forward, breaking the line, only to be sharply rebuffed with heavy loss by Greene's Continentals; Greene was not slow to take advantage of their disarray to order a general charge. There was some very severe bayonet fighting, but the British line held on the extreme right, and Washington's attempt to pass his cavalry between the Creek and Marjoribanks's men ended with Washington being wounded and taken prisoner. Gradually, the British were pressed back into the area of the house and garden. Marjoribanks held the right flank well anchored while Major Sheridan had by now occupied the house and his men poured a most destructive fire from its windows on the gun teams of the four six-pounders that the Americans had hurriedly brought forward.

---

*Stedman (p378) puts the number at 400, but this must be wrong, for it would be almost half Stewart's entire force.

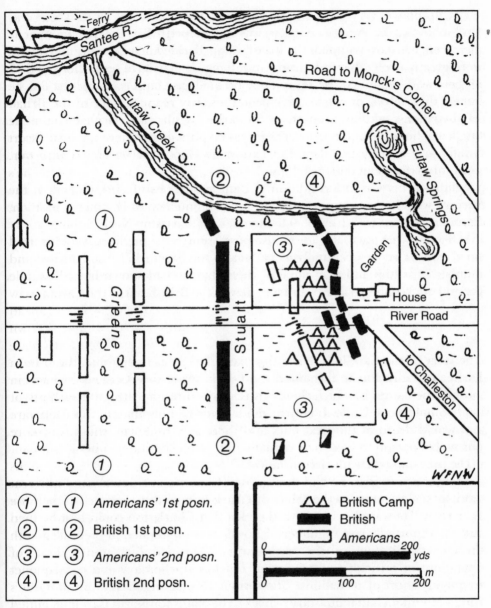

Ferry
Santee R.
Road to Monck's Corner
Eutaw Creek
Eutaw Springs
N
① ② ④
③
Garden
Greene
Stuart
House
River Road
③
to Charleston
② ④
① WFNW

① -- ① Americans' 1st posn.
② -- ② British 1st posn.
③ -- ③ Americans' 2nd posn.
④ -- ④ British 2nd posn.

△△ British Camp
� British
☐ Americans

0 _____ 200
yds
0 _____ 100 _____ 200
m

The Battle of Eutaw Springs, 8 September 1781

Confusion was very great and there was no telling how the battle would end. Undoubtedly, the American cause was not helped when a large number of their men, while overrunning the British camp, decided that drinking and looting were greatly preferable to fighting and dying. It may have been this, for their officers who continued the fight alone became easy targets, as indeed did the looters themselves, or it may have been the tardy reappearance of the British left battalions, adding strength to Stewart's assault, that decided Greene to break off the battle and retire.* There was no pursuit, for there was insufficient cavalry and a fairly pointless, but extremely bloody, battle ended with both sides suffering about 600 casualties.

Neither army was in a condition to fight another battle and, indeed, Eutaw Springs was the last engagement of any consequence to be fought in South Carolina. Stewert withdrew on 9 September to Charleston Neck, hoping at least to safeguard that town, the last British possession in the province, save for a few small posts. Greene eventually went back to the Santee hills: he had fought a number of battles and won none of them, but he had shown himself to be a general of exceptional ability and had driven the British out of the Carolinas.

*   *   *

The story of the Carolinas in the first part of 1781 is, in epitome, the story of Earl Cornwallis, for he fashioned the policy, made the decisions and fought most of the battles by which the British hoped to restore the royal authority in the two provinces. Undoubtedly he made some bad mistakes in this campaign, and would do so in the next one, and there are those who think he was an unsuitable person to have had virtually untrammelled independent command in the prevailing difficult circumstances.

From the very first, he had disregarded Clinton's instructions and advice to pay especial attention to the safety of Charleston, and to operate up the Cape Fear river. He was obsessed with the idea that Charleston could best be held, and the reportedly large number of Loyalists strengthened in their allegiance, through pacification of the backcountry. This led him into a series of operations in difficult country out of touch with the navy and consequently frequently short of provisions. The serious reverse in the autumn of 1780 at King's Mountain had temporarily checked his plans for North Carolina, but the arrival of General Leslie's troops, combined with news of Brigadier Arnold's expedition to the Chesapeake, revived them, for he had always pressed Clinton for diversionary assistance in Virginia.

The immediate result of this improvement in his affairs was Cowpens, for which Cornwallis can bear no blame but, undeterred by what he called 'a

---

*Strangely Stedman, who was not at this battle but was a very painstaking contemporary historian, makes no mention of this incident that has received prominence in almost every other account of Eutaw Springs.

diversion made by the enemy towards Ninety-Six', in which Tarleton lost him most of his light troops, he started on that disastrous march that took him along the edge of the Blue Ridge Mountains to the borders of Virginia and back again to the mouth of the Cape Fear. The letter he sent to Clinton of 18 January 1781[23] in which he reported the Cowpens battle, and told Clinton he was beginning his march, was the last communication he had with the Commander-in-Chief until he wrote from Wilmington in April. As Clinton did not receive this dispatch from Cornwallis until May, he was without any news from his subordinate for over three months, which had a serious effect on his planning.

Cornwallis's reasons for crossing the Yadkin have already been mentioned. The truth is that by then he had got himself into a situation whereby he had to defeat Greene if the latter was to be stopped from turning all potential Loyalists against the Crown. This he made clear in the first of two dispatches he sent on 17 March to Germain.[24] The second contained a glowing account of Guilford Courthouse, which of course gave no indication to the Colonial Secretary that the result of that battle might possibly herald the loss of the Carolinas. However, after Guilford and quite probably some time before, Cornwallis realised that the loyalist cause in the Carolinas was lost and, in a further dispatch to Germain of 18 April, he was to say as much.[25] His opinion thus expressed did little to alter Germain's firm belief in the continuing value of loyalist support, but it gave the Opposition another prop in their attempt to persuade the government that the war was lost.

After Guilford, Cornwallis withdrew to Cross Creek, then to Wilmington and then to Virginia along the stony path that led to Yorktown. Both at Cross Creek and at Wilmington he had momentous decisions to make in regard to the safety of Charleston, but it has been shown that he was quite convinced that there were sufficient troops in the province for that purpose without his small army. In fact, Charleston was in considerable danger, and Rawdon had far too few troops for his many commitments. Cornwallis would have done much better to have joined Rawdon at Camden rather than pursue his cherished hope of partaking in Clinton's Virginia campaign, the strategy of which he never fully understood, and had undermined when he ordered General Leslie into South Carolina. However, he decided to march north and, although he had been out of touch with the commander-in-chief for three months, he did not even bother to wait in Wilmington for the arrival from Charleston of three dispatches from him that he knew Colonel Balfour was forwarding; not that these, probably, would have altered his decision.

And so the Carolinas were lost, and the blame must be attributed almost entirely to Cornwallis. Cowpens was a disaster which Cornwallis regarded as a personal affront, and he was determined to avenge it, and speedily. This resulted in an unnecessary destruction of baggage and supplies, with consequent misery and deprivation for his army, and an improperly planned advance over extremely difficult country. Greene was an excellent general, and he held most of the cards, but Cornwallis was capable of defeating him without ruining his army in the process had he taken it more slowly, and perhaps made greater use of Rawdon. The decision at the end to forsake the province and march north can be understood, but not condoned, for it was in flagrant disobedience of orders to safeguard Charleston and South Carolina.

Cornwallis had proved himself a very good general before, and was to do so many times again. He was a brave and greatly respected leader, and it could be said that he was the only general on the British side in the entire war whose tactics in battle were almost invariably sound. But in the Carolinas he had shown himself to be a man who disliked bowing to the dictates of others, or even to the force of events, and at times irritation and impatience led him to adopt a self-centred course that conflicted with the responsibilities that had been thrust upon him.

CHAPTER 13

The Last Campaign

THE LAST LAND campaign of the war was fought in Virginia. It was muddled from the beginning by a mass of conflicting correspondence between Clinton, Cornwallis and Germain, and between Clinton and the two generals Arnold and Phillips, who were already operating in Virginia. In the end, it was doomed through the inability of the British navy to retain command of the sea in American waters.

Back in August 1780, Clinton had put forward to Germain a plan, originating from Colonel William Rankin (Colonel of British Militia), that showed the strategic advantages of an operation against the Delaware Neck,[1] that piece of land bordered by the Chesapeake and Delaware Bays. Later, his thoughts were to combine this with a strike north up the Delaware against Philadelphia in conjunction with troops moving down from New York. This had received approval in London, but the extra 4,000 troops required for it were not readily forthcoming and Cornwallis's withdrawal of Leslie's force for his North Carolina campaign had put it temporarily in abeyance. However, it was never far from Clinton's mind, and in the spring of 1781 was to be revived by Clinton and criticised by Cornwallis. The latter was convinced that until Virginia was subdued the Carolinas could never be held, and therefore its subjugation should have priority.[2] In this he was supported by Germain,[3] despite the latter's earlier approval of Clinton's plan.

In January 1781, Clinton had sent Arnold to Virginia in response to Cornwallis's request for a diversion, and to establish a post at Portsmouth or elsewhere capable of protecting the King's ships and maintaining a garrison of 500 or 600 men. At the end of March, General Phillips arrived in Chesapeake Bay from New York with a further 2,000 men. In the course of various conversations that Clinton had with Phillips before he sailed, and in a letter to him of 26 April, he made it very clear that while Phillips should give priority to assisting Cornwallis in securing the Carolinas (Cornwallis's letter to Clinton of 10 April had not mentioned his decision to march north almost immediately into Virginia) he was to be prepared for the possibility of implementing Clinton's Chesapeake-Delaware operation. He also gave Phillips his thoughts for the best place for the protection of the ships, saying 'I know of no place so proper as York Town', or if that proved impossible he recommended Old Point Comfort.[4]

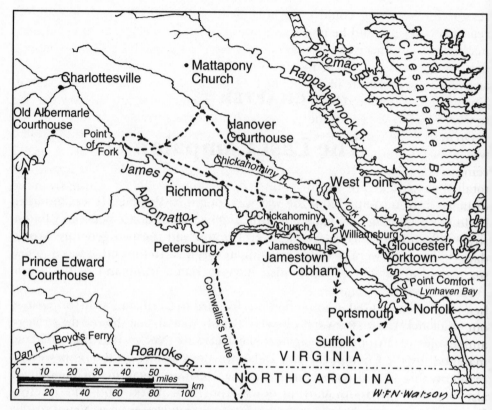

Cornwallis's Campaign in Virginia

It was a great loss to the army when Phillips died of a fever on 13 May at Petersburg, where he had gone in response to a letter from Cornwallis, written the day before he left Wilmington. Neither Phillips nor Arnold had had any significant contact with the American forces during their operations in the Williamsburg peninsula, but in their raids up and down the James and Chickahominy rivers to Richmond and beyond, they had done immense damage to stores and installations. They had also fortified Portsmouth, although that was later to be considered an unsatisfactory base. While always hoping to use Phillip's men for his intended operation up the Delaware, it was not easy for Clinton to make a realistic plan while he was totally unaware of Cornwallis's movements. When the latter arrived at Petersburg on 20 May, he read Clinton's correspondence with Phillips and immediately formed strong reservations concerning Clinton's proposed advance into the Middle Colonies, greatly preferring to seek out Lafayette's army in Richmond.

Clinton was, quite rightly, very angry with Cornwallis for abandoning the Carolinas and marching into Virginia without any authority. His strategic reason for so doing could only hold good if he had left South Carolina reasonably secure, and he must have known that Greene's continued presence made this most uncertain. Cornwallis was later to excuse his action by relating the dangers of the Carolinian summer climate to the existing sickness of his army, but it seems more likely that he looked forward to fighting a campaign in Virginia unencumbered by the distraction of having constantly to show the flag to would-be Loyalists – for there were no longer many in that province.

On arrival at Petersburg, Cornwallis took over command from Arnold (who was soon to return to New York), and with the addition of a recent reinforcement from New York of some 1,700 men, he had an army that totalled nearly 7,000 of which about 5,000 were fit for duty. Clinton, from the distance of New York, could do little more than point out to Cornwallis, when he learnt of his intention to go after Lafayette, what he considered to be the strategic advantages of his Middle Colonies plan, and to warn him of the dangers of operating in the Chesapeake should the navy lose command of the sea.

However, this monition had not yet been sent, and in crossing the James and heading for Richmond to engage Lafayette, Cornwallis would only be carrying out what had been Phillips's intention before he received Cornwallis's letter announcing his imminent arrival at Petersburg. And so, on 26 May, he marched. The Marquis de Lafayette had come to Virginia in March with orders primarily to catch the traitor Arnold, whom the Americans dearly wanted, but also to create a diversion to assist Greene in South Carolina. This young Frenchman was only 23 and full of enthusiasm and, although no Napoleon, he possessed a certain degree of tactical skill. He did not wait in Richmond for Cornwallis, but crossed the Chickahominy and struck north to join up with Anthony Wayne. There followed some cat and mouse tactics in which no contact was made, and Cornwallis, tiring of the chase, fell back to Williamsburg. He did, however, detach Tarleton's British Legion and another irregular force, the Queen's Rangers, to raid in the areas of Point of Fork and Charlottesville. The Queen's Rangers had come to Virginia with Arnold and were commanded by Lietenent Colonel John Simcoe, who had all of Tarleton's courage and brio, but was less of a maverick and more of a regular soldier. These two forces caused very considerable damage to arsenals and store houses, and Tarleton's men narrowly missed capturing Thomas Jefferson when they irrupted into the Virginia Assembly in Charlottesville.

On his arrival at Williamsburg, Cornwallis found two letters from Clinton which were to be the first of a whole series to come in the course of the next few weeks that could do nothing but muddle, confuse and irritate him. These two, written on the 11th and 15th of June[5] and arriving on the 26th, pointed out yet again the advantages of the Delaware Neck plan, recommended that he should establish a defensive post at Williamsburg or Yorktown, and that he should then be prepared to send Clinton all troops not wanted for the defence of that post. (Throughout this correspondence Clinton was inclined to recommend or suggest rather than to order, which added to Cornwallis's difficulties. Clinton was later to say that he did this because he understood the government greatly preferred Cornwallis to him, and he expected to be relieved of his command at any time.) Letters had been intercepted that showed that Washington was preparing to lay siege to New York with at least 20,000 men, and Clinton – juggling with figures somewhat – reckoned that Cornwallis had considerable numerical advantage over Lafayette, and would have troops to spare. Cornwallis was not pleased with these letters, for he could see no possible point in reducing the Virginian expedition to the occupation of a single post, and in this he was right, although Clinton was also right in that it was very important to have a secure base from which the navy could operate.

However, the recommendation to find a port, and the expressed wish to send troops, had to be obeyed and Cornwallis took Simcoe on an extensive but fruitless reconnaissance. They considered Yorktown could not be held after sending to Clinton the troops he had specified, and one or two other places were also discarded. It appears that they misunderstood Clinton's dispatch, although it seems fairly clear, in that they were only to send Clinton those troops that could be spared after keeping sufficient for the security of the base. Back at Williamsburg by 30 June, Cornwallis wrote to Clinton that day a long letter starting with an account of the two raids, and the successful little rearguard action Simcoe had fought with Lafayette when the army entered Williamsburg. He then again touched upon the advantages of a Virginian campaign over the Delaware one, and gave his reasons for turning down Yorktown as a base. The letter concluded with his opinion that there was little chance of being able to establish a post capable of being able to protect ships of war, and 'as I do not think it possible to render any Service in a defensive situation here, I am willing to repair to Charleston'.[6] Cornwallis must have known that it was most unlikely that Clinton would agree to his returning to Charleston at this stage of the campaign, and the offer was probably made because he was losing patience with the whole business.

On 4 July, he decided to march his army from Williamsburg across the James to Portsmouth, where the troops for Clinton would be embarked, but an equally important motive may have been to lure Lafayette into battle, the latter's army being encamped only a few miles from Williamsburg. The river was to be crossed by Jamestown Island with the help of a naval detachment under Captain Aplin. Cornwallis sent the Queen's Rangers over on the evening of the 4th to cover the baggage and carts which were to cross during the next two days. Meanwhile, the rest of the army took up a strong natural position covered by ponds on the right, and in the centre and left by a morass stretching down to the river with a few narrow causeways connecting it to the strongly wooded countryside. On his left, Cornwallis placed Dundas's brigade of three regiments (43rd, 76th and 80th) and two six-pounders. On Dundas's right was Yorke's brigade, the Guards' battalions and then the Hessians. The Legion's cavalry formed a second line in rear of the 80th Regiment. The total force was around 4,800 men and Lafayette's army, encamped near Chickahominy Church, was of about the same strength, but only 1,500 of his troops were Continentals.

On the morning of the 6th, cavalry patrols reported the approach of the Americans with Anthony Wayne's Pennsylvanians acting as advance guard. The Pennsylvania Line had mutinied a short while back over pay and conditions, but that was over and they were very good troops, their commander had much battle experience and invariably led from the front. He now expected a fairly easy fight against Cornwallis's rearguard, the only troops not yet across the river, and he was confirmed in this by two 'deserters' – a negro and a dragoon – whom Tarleton had briefed to join the Americans to report that only a detachment of the British Legion still remained on the left bank of the river. Tarleton's cavalry was ordered to support the pickets against Wayne's leading troops while the main army remained quiet and hidden from view. The ruse worked splendidly, and Wayne's men crossed the bog sublimely ignorant of the trap they were entering.

The dragoons fell back through the ranks as Cornwallis advanced the army against the badly surprised, but in no way daunted Wayne, whose men fought with great gallantry until Lafayette hastened up on the left with some 900 Continentals and 600 militia. Here the fighting was at its heaviest in a close-quarter battle of infantry and artillery; men were bayoneted, and blasted from almost point-blank range, while horses could be seen stumbling from the field, blood streaming from tremendous gashes. Lafayette himself had two horses shot under him, and his gun teams suffered similarly and lost two guns. On the British right, the fighting was comparatively light, which enabled some of Yorke's men to swing in against the American left. The battle had not started until around sunset, and was ended by darkness, with the Americans retreating towards Green Spring Farm which was to give its name to this short, but severe and largely unsung battle. Tarleton felt strongly that Cornwallis had missed an opportunity to destroy Lafayette's army by not pursuing,[7] but Cornwallis rightly felt that the need to get the troops to Portsmouth for embarkations should have priority.

After the battle, therefore, while Lafayette withdrew northwards, Cornwallis continued his crossing of the James and headed for Suffolk County. The army camped briefly at Cobham, and from there, on 9 July, Tarleton was sent on a raid towards Prince Edward Courthouse to destroy stores destined for Greene's army. However, the stores had already been dispatched and the expedition was a rather costly failure. At this time, Cornwallis received a letter from Clinton written on 28 June which made no mention of the threat to New York, but talked instead about a rapid move against Philadelphia to destroy stores there, for which he wanted Cornwallis to send him a number of regiments, which he again specified, if Cornwallis 'had not already embarked the reinforcement I called for in my letters of 8th, 11th, 15th and 19th instant'.[8] As only one of those four letters had at that time been received by Cornwallis, this would seem the place to give a brief outline of Clinton's confusing and contradictory correspondence which clearly illustrates the difficulties under which Cornwallis laboured.

On the army's arrival at Suffolk on 12 July, he received three letters from Clinton written on 19 May, 8 and 19 June. The one written in May was long and rambling, being concluded on 1 June, and mainly concerned Clinton's displeasure with the removal of the army from the Carolinas,[9] but that of 8 June was sharper and more peremptory in tone. It was the precursor of the one written three days later, quoted above, which had arrived a fortnight earlier giving news of the intercepted enemy letters, one of which from Layfayette clearly showed, according to Clinton, that Cornwallis should be able to spare him 2,000 troops. The letter went on to say that now Cornwallis was so close there was no longer any need for him to send dispatches to Germain; 'you will therefore be so good to send them to me in the future.'[10] The records seem to show that his request was complied with.

The letter of 19 June was altogether more conciliatory, but Cornwallis must have found it rather muddling. The threat to New York, somewhat mysteriously, was already beginning to take a back place, and therefore Clinton would not press for the reinforcements requested, 'But if, in the approaching inclement season, your Lordship should not think it prudent to undertake operation with the troops you have, I cannot but wish . . . you would send me as soon as possible what you can spare from a respectable defensive.' A little later in the letter, the tempo is changed to 'In the hope that your Lordship will be able to spare me three thousand men, I have sent two thousand tons of transports from hence.'[11] These two June letters only partially tied in with the dispatch Cornwallis had received four days earlier about the Philadelphia raid.

However, as the most persistent thread running through this correspondence was the need for troops in varying numbers, Cornwallis sent them to General Leslie, commanding at Portsmouth, and by 20 July Leslie's preparations were well forward and the Queen's Rangers had actually embarked. Then at 1am on the 23rd, Brigade Major Bower arrived bearing a brief dispatch (dated 11 July, and to be followed the next day by a fuller explanation cancelling the whole movement, and intimating that Cornwallis's troops were to be used for the defence of Old Point Comfort or Yorktown.[12] It is small wonder that Cornwallis was said to have been exceedingly angry on receiving this message which overturned all his arrangements and confirmed the June letter that his highly professional army was to be reduced to the seemingly humble and what he considered unnecessary, role of defending a port.

Cornwallis remained a few days at Portsmouth awaiting Tarleton's return during which time he might perhaps have pondered on this conflicting correspondence that emanated from New York with such tiresome frequency and in particular the principal event that started it all – the threat to New York. This was in fact very real, and had been made quite clear to Clinton by an intercepted letter from Washington to Lafayette dated 31 May informing the latter that he and Generals Rochambeau and Chastellux had decided, at a conference held at Wethersfield, Connecticut, on 21 May, to combine in an attack upon New York.[13]

Lieutenant Général Comte de Rochambeau, who had arrived to take command of the French troops in America almost exactly a year before, had very soon seen that the Americans could win the war only with active French support, which hitherto had not been very effective, and he had been largely responsible for the dispatch to Caribbean waters in the spring of 1781 of Lieutenant Général Comte de Grasse with 26 sail of the line and several thousand troops.* If Admiral Sir George Rodney had allowed his very capable second-in-command, Admiral Sir Samuel Hood, to station his 18 ships of the line to windward (east) of Martinique as he wished, instead of to leeward as Rodney ordered, de Grasse might not have brought his whole convoy safely to port. Hood, cruising to leeward west of Martinique was in the worst possible position to intercept the French fleet approaching from the east, and to engage it closely. As it was, de Grasse, manoeuvring with very considerable tactical skill, held to a southerly course so as to shield his transports while they hugged the coast up to Fort Royal. Then, keeping his distance, he engaged Hood's fleet in some very damaging long-range gunnery.

Immediately after the Wethersfield meeting, Rochambeau had written to de Grasse requesting him to bring his fleet so that, in conjunction with Admiral Barras, they would be able to close New York port. But preparations for the land attack began at once with the concentration of American and French forces from Peekskill and Rhode Island respectively in the area of Dobbs Ferry-Sawmill River-White Plains. There were skirmishes with the British outposts which were withdrawn into the main New York lines, and Generals Lincoln (who had been exchanged for General Phillips a few months after the surrender of Charleston) and de Chastellux carried out a reconnaissance in force with some 5,000 men along the whole British front.

It was this that prompted the second batch of Clinton's letters to Cornwallis, for it must have seemed certain that an attack was imminent. However, the arrival in New York of 2,000 Hessians at the beginning of August obviously went a good way to relieve Clinton's anxiety for, on 16 August, he wrote to Admiral Graves, who had recently replaced Arbuthnot, suggesting that these Hessians could not be better employed than in an attack on Rhode Island before that place was strengthened by de Grasse. The captured correspondence had indicated de Grasse's likely arrival, and it seemed a wise move to pre-empt him, but Graves answered that owing to necessary repairs he had not the ships available.[14]

As it happened, there was to be no attack on New York. The fortifications

---

*De Grasse was not, as was sometimes the case, an army general given charge of a fleet with substantial troop transports, but a lieutenant general of the naval force of France, and an officer with considerable experience of naval warfare.

were strong, the British had been reinforced and both Washington and
Rochambeau were coming to the conclusion that their combined army would
be better employed against Cornwallis in Virginia. Their minds were resolved
when the frigate *Concorde* arrived in Newport with a message that de Grasse
would be leaving St Domingo on 3 August for the Chesapeake with his entire
fleet and some 4,000 men, but he must be back in Caribbean waters by
mid-October. In great secrecy, preparations were immediately begun to march
the army across New Jersey and Pennsylvania to Virginia, and very successful
deceptive measures were thought out. It was necessary to leave General Heath
to safeguard the Hudson posts, which meant that when Washington crossed
that river at King's Ferry on 21 August he had only 2,000 men, but the French
army of nearly 5,000 strong joined him on the 25th. It was not until 2
September that Clinton learnt for certain what Washington was up to.[15] On
that day, the allied army was receiving a rapturous welcome as it marched
through the streets of Philadelphia. It had been a beautifully planned and
executed operation, and was perhaps Washington's finest hour of the war.

It must have been very clear to Clinton that, more than ever now, the
outcome of the war depended upon command of the sea, and that the British
in North America could no longer be certain of it. At the beginning of 1781
the British had 32 ships of the line in the West Indies and eight in North
America against 21 French and Spanish ships in the West Indies, and seven
French ships in North America. This advantage in numbers was not to last, for
on 28 April de Grasse sailed into Fort Royal with his large fleet, while British
ships were frequently required for convoy duties. During the summer and
autumn the balance became heavily tilted towards the allies, which enabled
them to take the important Florida port of Pensacola in May, [16] and at the time
of Yorktown the French and Spanish could muster an overall total of 45 ships
of the line as opposed to 31 available to the British.

True to the message he had sent to Rochambeau, de Grasse sailed for the
Chesapeake on 5 August with 27 ships and 3,200 troops under the Marquis de
St Simon.[17] This was his entire squadron, and it left the French islands and
their trade very vulnerable, and was a move so bold and unexpected that
Admiral Rodney, who had been ordered to keep a close watch on de Grasse
was taken unawares. The Admiral was unwell, and at the beginning of August
he handed over his command to Admiral Hood and sailed for home – in a
convoy escorted by no fewer than three ships of the line and two frigates.[18]

Rodney was a great sailor (if at times a difficult and choleric man) who was
soon to distinguish himself in these very waters, and the mistakes he made at
this time were probably due to his illness. He had learnt on 31 July that a frigate
from Newport had brought 30 pilots for the Chesapeake and Delaware,[19]
which made him fairly certain of the French fleet's pending destination, but he
failed to notify Graves of this until it was too late to be of use, and he did not
equate the number of pilots with the likely strength that de Grasse would
deploy. He did, however, stress in a letter to Sir Peter Parker, commanding at

Jamaica, that as the enemy had a very strong naval force 'I must beg you will dispatch the *Torbay* and *Prince William* without detaining them a single moment, and also that you will add to their force by sending . . . every line of battleship you can possibly spare from your station.'[20] Of the two commands, Jamaica was usually considered more important than the Leeward Islands, and Rodney probably felt that 'beg you' was as positive as he should go, and that it would produce enough ships for the purpose, especially as he still thought de Grasse would divide his squadron. However, it did not convey sufficient urgency for Parker to send more than the two ships specified, and they arrived too late. His letter to Lord Sandwich of 1 September seems to suggest that a larger contribution from him was unnecessary.[21]

As soon as he had learnt of de Grasse's intended destination, Rodney ordered Hood to sail north and, at Clinton's request, he embarked a regiment of foot. He sailed on 10 August with 14 ships of the line and arrived at Chesapeake Bay just before de Grasse (who had sailed five days earlier). Finding no sign of the French fleet there or in Delaware Bay, he sailed on to New York arriving on 28 August.* There he reported to Admiral Graves who had only five ships of the line fit for service, but learning that Comte de Barras had left Newport for the Chesapeake, Graves took command of the combined fleet of 19 ships and sailed on 31 August.

Graves reached the Chesapeake on 5 September without knowing exactly how many ships he would have opposed to him, but he had hoped to intercept de Barras's squadron before it could join de Grasse's, whose ships definitely were by now anchored in Lynhaven Bay with a few still up the James. As soon as the Frenchman realised he had the enemy approaching, and not de Barras's expected squadron, he up anchored and sailed out of the Bay in a southerly direction with what ships he had, leaving the rest to follow. There was a muddle in the British fleet over the signal for the line of battle and the signal for battle. This came about through the signals used by Hood's West Indies ships being different to those of Graves's North American ones, and there had not been time to discuss this before the combined squadrons sailed. This, together with Graves's lack of information and his limited acquaintance with most of his captains, prompted undue caution through which he missed an opportunity to attack the French at a disadvantage while they were beating out of Lynhaven Bay through a narrow channel.[22] This was a mistake that had serious consequences.

There followed several days of manoeuvre in the open sea without a serious battle, but a number of sharp, although indecisive, engagements. The damage to ships and loss of men on both sides were fairly heavy, the British having 336

---

*Fortescue (p392) says that de Grasse was already there anchored in Lynhaven Bay, which owing to the timings would seem quite probable. This is not borne out by reports and dispatches, and it is certainly strange that Hood could miss 23 ships at anchor. The other four were said to be unloading troops up the James.

casualties and the French 220 men killed and wounded.[23] Hood was to argue later that while de Grasse was still at sea, Graves should have slipped into the Bay and anchored in line of battle at the mouth of the York between the French fleet and the British defences. It probably was not a practical course of action and, in any case, during the fighting de Barras had beaten Graves to it and sailed into the Bay with seven ships of the line and Washington's heavy guns and stores. Graves did enter the Bay after the battle, but on learning that de Barras was higher up, he decided, in consultation with Hood, to return to New York for repairs and to replenish his rapidly diminishing stores.

At the time of these naval engagements, Cornwallis's army was busy preparing the defences of Yorktown. There had been two further letters from Clinton on the matter of bases written on 8 and 11 July, brought together by the hand of Captain Stapleton. That of the 11th was the follow-up to the brief dispatch brought by Major Bower cancelling the order to send troops to New York, which had made Cornwallis so angry. In this letter, Clinton said that he and Graves favoured Old Point Comfort, but he suggested that Yorktown should also be occupied as security for the works at Old Point Comfort. The question of numbers for each post was left to Cornwallis's discretion, and he was to keep all the troops he wanted for them.[24]

On receipt of these letters, Cornwallis had sent his engineer, Lieutenant Sutherland, to reconnoitre Old Point Comfort, but he was back in a very few days reporting that he found the ground totally unsuitable for a defensive post to protect ships. Cornwallis then went himself with some ships' captains, all of whom agreed with Sutherland. As a result, it was decided to take possession of Yorktown and Gloucester using the whole army, for in both bases considerable fortifications would need to be built. At the beginning of August, the army had embarked in what transports were available and sailed for Yorktown, General O'Hara being left with a party to destroy the fortifications at Portsmouth. Leslie had been sent to command at Charleston and rejoined the army three weeks later.

Cornwallis had begun to fortify Gloucester as soon as the army had assembled on the peninsula, but work was not begun on Yorktown until later. This formed one of Clinton's numerous post-war condemnations of Cornwallis,[25] for Yorktown was the commanding feature and was important for Gloucester's security. Cornwallis, in his rebuttal of Clinton's accusations, makes no mention of this criticism, which only later assumed importance, and even then from the timing standpoint and not the tactical, so he probably thought little of it. His mind at the time was more likely to have been on what became one of the great controversies of this most controversial war. Should he, and could he, have broken out of the trap that was rapidly closing upon him?

When the British first occupied the two posts, Lafayette had only a small army in the area of the William and Mary College at Williamsburg, but when de Grasse had landed St Simon's marines at the beginning of September, the combined Franco-American army was about 5,600 men with a substantial backing of militia in reserve. Cornwallis's strength return for 15 August showed 7,500 total effectives of which 5,958 men were fit for duty.[26] So he was quite substantially outnumbered, although, until Washington's army arrived three weeks later, he still had a reasonable chance of escaping.

As soon as Clinton had discovered that Washington had deceived him over Philadelphia he had written to Cornwallis on 2 and 6 September informing him that Washington was by then probably at Trenton with some 6,000 troops.[27] The first of these letters arrived on 15 September, and while Cornwallis probably had some previous knowledge of Washington's march, it is unlikely that he knew of its progress or of its size. Largely because Clinton had repeatedly assured him that both he and the navy would come to his support (although Cornwallis probably knew more than did Clinton about the naval situation), Cornwallis decided against any attempt to break out before being closely invested. It was a crucial and difficult decision to take, and the reasoning that prompted it will be examined shortly.

While Cornwallis was pondering the advisability of a break-out, the defence works in the two posts were being hastened along and Clinton was making his own efforts to ease his deputy's plight. Cooped up in New York, Clinton had become isolated from the main theatre of events; he was out of touch with the up-to-date naval position and, although Germain had constantly assured him that Rodney would take care of the French squadrons, this had not proved to be the case. He saw clearly that Cornwallis was in serious trouble and fully intended to go south himself with what troops he could spare from New York when he got naval clearance that the coast was safe. Meanwhile, he sent Arnold on a large-scale raid into Connecticut in what one supposes he thought would create a useful diversion. Two forts were carried by assault with heavy casualties, New London together with a number of ships and a great quantity of stores were destroyed, and Arnold and his men were guilty of excessive cruelty. As a diversion, the raid was a total failure, as Clinton must have known it would be had he given it proper thought.

While Arnold was bringing this further obloquy on himself, Washington's troops were sailing down the Delaware from Trenton to Christian Creek, and the Yorktown defences were not yet ready to receive them. It is so easy from a distance in time, and without full knowledge of prevailing conditions, to be critical, but it would seem that the discarding of Old Point Comfort was a serious mistake. Both the army and navy would have had more room to manoeuvre there, whereas at Gloucester and Yorktown both were bottled up. Sutherland's dislike of Old Point Comfort seems to have been based on a shortage of suitable materials for earthworks and redoubts and possibly, though not probably, that was enough to rule the place out. Anyway, Cornwallis seemed happy enough with the chosen site, although, of course, at the time he did not know that the navy would be unable to support him. If Yorktown was to be the main base then it was necessary to fortify Gloucester in order to protect the ships and this, as we know, Cornwallis had begun to do as soon as he had had the army assembled in the peninsula.

Gloucester was no more than a hamlet and, apart from swamps on both flanks, there were no natural features across the open expanse of flat ground to lend strength to artificial ones. Cornwallis had sent over the Queen's Rangers, the North Carolina Volunteers, the light companies of the 23rd and 82nd and some Hessians,[28] and these were joined by Tarleton's cavalry on 2 October. They were opposed by the Duc de Lauzun's cavalry, some Virginian militia under General Weedon, and 800 marines from de Grasse's ships.[29] Simcoe's men carried out a number of successful foraging expeditions, but the only serious engagement on this side of the river involved Tarleton's dragoons in combat with de Lauzun's hussars. The action was hard fought, and Tarleton was unhorsed but escaped capture through seizing a riderless mount. The dragoons, whose cavalry drill was slipshod, were outnumbered and forced to retreat to the protection of their infantry which did considerable damage to de Lauzun's pursuing troopers. The British lost one officer and 11 men, and the French two officers and 14 men.[30] It was Tarleton's last engagement of the war, and conducted with his usual élan.

Yorktown offered Cornwallis a much greater degree of natural protection. His defensive line was semicircular and each flank rested on the river. Additionally, it was protected by two creeks, on the right of the line the Yorktown Creek and on the left the Wormley Creek. In front of the Yorktown Creek there was a deep ravine that covered about half of Cornwallis's front, but there was a vulnerable area between the two creeks where there was no ravine. Here Cornwallis built three redoubts and these formed part of his outer line of defence, which was separated from the inner line by almost half a mile. Other smaller fortifications were constructed on the outer line at the head of the Wormley Creek, and a large redoubt, fronted by an abatis and supported by three batteries, was built on the extreme right a little way in front of the Yorktown Creek. This was known as Fusiliers' Redoubt. Two frigates lay at anchor at the mouth of the river, and on 16 September 10 transport ships were sunk between Gloucester and Yorktown.

There were a number of roads leading to the position along which the allied army was to advance, but the terrain dictated that the major assault would come in on the left of Cornwallis's line. At the end of September, when the siege began, the British had 65 guns in position, but the total manpower fit for duty had fallen to 5,129 with a possible call on 800 sailors or marines. The combined allied army was in the region of 16,000 men, half of whom were French. The American half was divided between 5,000 Continentals and 3,000 militia.[31] In a strongly defended position, adequately supplied, three to one against are acceptable odds, but at Yorktown these important prerequisites were not met.

Washington and his staff had arrived with Lafayette a few days in advance of his troops, but his whole force was concentrated at Williamsburg by 26 September. Two days later, the allies began their investment of the Yorktown

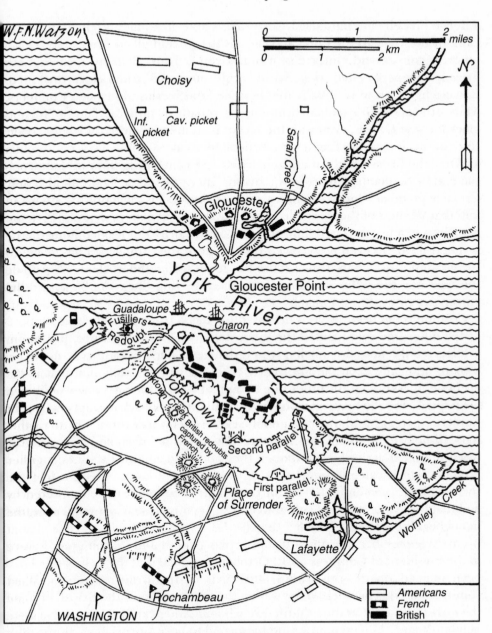

The Siege of Yorktown, 1781

defences. Their line paralleled Cornwallis's and, like his, stretched from one bank of the river to the other so that the British and allied forces faced each other in convex and concave semicircles respectively. Broadly speaking, the French troops fronted the British right and the Americans their left. Cornwallis's men were thus firmly held on land by this vastly superior force, while at sea de Grasse had brought his squadron into the Bay to form a blockade that Graves, even with the recent increment of Admiral Digby's three ships of the line, would find it very difficult to break.

While the French and American troops were completing their investment at the end of September, there was a drastic alteration made to the British line. On 29 September, Cornwallis had received Clinton's dispatch of the 24th which said that 23 ships of the line with 'above five thousand men rank and file' would be sailing 'in a few days to relieve you'.[32] To this Cornwallis replied immediately saying 'I have this evening received your letter of the 24th, which has given me the greatest satisfaction. I shall retire this night within the works . . .'[33] And on the next day the outer line was abandoned, leaving the redoubts and other entrenchments to be occupied by the allies.

Cornwallis has been severely criticised for this retrograde step, because parts of the outer line commanded the inner defences. In consequence, the enemy artillery could now do much greater damage. The decision was made, of course, upon the expectation of early relief. Cornwallis reckoned that in the event of an assault on the outer line by a greatly superior force, he would have been sacrificing the forward troops, who could not fall back in safety across the numerous creaks and ravines under severe enemy pressure. On the other hand, if he brought them back at once, they would be available for a break-out when the relief arrived, even though they added to an already very congested position, which extended only to 1,200 yards by 500 deep. It is possible that there had not been enough time for the troops to complete the outer defences, and that they were still very vulnerable. If so, it emphasises Cornwallis's mistake in not beginning the work earlier than he did.

On 30 September, the allied troops broke ground and Washington, writing to the President of Congress on 12 October, reported that the first parallel had been opened on the 6th within 600 yards of the British defences. Work continued on this parallel, which was on the extreme left of the British line, for the next two days. On the 9th, the French and American batteries opened the bombardment.[34] Lieutenant Ewald happened to be at Cornwallis's dining table when the first shot was fired, and his diary reports, 'By the first cannon shot of the besiegers Commissionary Perkins was killed at table and Lieutenant Robertson of the 76th Regiment lost his left leg. The lady of the good Commissary General sat between the two at table during this misadventure.'[35] The poor lady must have had quite a shock, with worse to follow, for it was calculated that on that day the allied batteries fired some 3,600 cannon balls into the British lines.[36]

Thereafter the cannonade continued unabated, more guns being assembled alongside pyramids of cannon balls and sacks of powder. Buildings were shorn away under this steel flail of gunfire, the frigate *Charon* was set ablaze, and batteries were put out of action; but men of the 23rd still held fast in Fusiliers' Redoubt against the stream of shot and shell that was hurled against them. All this time, the Americans on the right were sapping forward from the first parallel with their zigzags, and making good use of the numerous shallow ravines, to prepare the second parallel. Cornwallis knew that, unless relief came quickly, his position could not be held. On 11 October, he wrote to Clinton 'nothing but a direct move to York river, which includes a successful naval action, can save me.'[37]

It was on that day that the second parallel, only 300 yards from the British defences, was opened, after what had been some fierce fighting in an assault by both American and French troops on two redoubts. These outposts had to be taken before the second parallel could be completed, and they were very strongly defended by Hessian and British troops. On the 15th, Cornwallis wrote again to Clinton saying, 'Last evening the enemy carried my advanced two redoubts on the left by storm, and during the night have included them in their second parallel . . . My situation now becomes very critical . . . The safety of the place is, therefore, so precarious, that I cannot recommend that the fleet and army should run great risk in endeavouring to save us.'[38]

Meanwhile, the situation in the town had become deplorable. There was little refuge from the shelling, and bodies lay about unburied. Food was desperately short and much of the meat was putrid, all of which necessitated driving slave labour from the town, many of the negroes being killed in the cross fire. Inevitably, the number of wounded increased daily and medical supplies for them became hopelessly inadequate. Forage was in even shorter supply than food. Hundreds of horses had to be slaughtered and their carcasses left to rot on the river banks. Cornwallis still hoped for relief, having received a letter from Clinton dated 30 September saying that he had reason to hope that the relief force 'may pass the bar by the 12th of October, if the winds permit, and no unforeseen accident happens'.[39]

These various dispatches Cornwallis was receiving originated from the deliberations of a council of war held in New York, attended by five general and four flag officers.[40] At first, only Major General Robertson, the military governor of New York, seemed to realise the urgent need for a strong relief force, but he did not press his case in opposition to Clinton and, with officers recently back from Yorktown giving a fairly optimistic report, it was decided to await the arrival of Admiral Digby, with between three and five additional ships, when, it was hoped, the navy might regain command of the Bay.[41] It was not until Cornwallis's dispatches of mid-September arrived, and alerted the council to his grave danger, that positive action was decided upon, although not without very considerable, and quite understandable, misgivings on the part of the admirals.

The council of war had been fumbling in the dark, for things were happening much faster in Yorktown than could be comprehended in New York. Cornwallis would have known that the date of 12 October, now given as the earliest for relief, was running it very fine. He must also have known that the convoy's chances of success were very slim. It was highly unlikely that de Grasse's captains were incompetent seamen, and Admirals Digby and Graves would have a severe problem in trying to get between Yorktown and the French fleet to land the relieving forces. By early October, the likelihood of surrender as the only outcome must have loomed large, but Cornwallis was still thinking positively. The first essential was to put those guns firing at close range from the second parallel out of action, and he thought that, as a last resort, a break-out from the Gloucester defences, which had not been subjected to any major attack, might prove possible.

A sortie was planned for the early hours of 16 October. Three hundred and fifty men, under the overall command of Lieutenant Colonel Abercromby, were ordered to attack the batteries at each end of the second parallel. Lieutenant Colonel Lake with a detachment of Guards and the grenadier company of the 80th Foot attacked one site, while Major Armstrong with a company of light infantry took on the other. Both batteries were covered by redoubts which had to be taken before the guns could be spiked. The assault was gallantly carried out, the redoubts were reduced, 11 guns were spiked and about 100 French and American soldiers were killed with little loss to the British. But the operation was fruitless, for the guns had to be spiked in a hurry and, being only half done, it was not long before the allied gunners had them back in action.[42]

By now, Cornwallis had very few of his own guns still firing, and his ammunition was nearly expended. The choices left to him were immediate surrender or to attempt the break-out, and then to strike north to join Clinton or perhaps south to Carolina. He chose the dangerous and difficult option of breaking out. Sufficient boats to ferry the army, less the wounded, artillery and baggage, were to be assembled by 10pm on the 16th, and Tarleton had been ordered to mount an attack from his defences early on the 17th. The first wave consisting of the light companies, most of the Guards, and some men of the 23rd got safely across the river without alerting the enemy, but no sooner had the second wave started than a semi-tropical storm, so common in those waters, raged up the river dispersing what boats were afloat, and making further crossings impossible. By the time it abated, in the early hours, daylight was too close. There was nothing for it but to send orders to Colonel Abercromby, who had charge of the first flight, to bring his men back. This proved no easy task (and in the circumstances one wonders why it was necessary). Boats could not be quickly collected after the storm, and de Choisy chose this time to launch his only major attack against the defences. Nevertheless, Abercromby's men were back with very few losses by midday.

Lieutenant Ewald, who spent much of the siege at Gloucester, commended in his diary Cornwallis's gallant attempt to break out, but emphasised the impossibility of success. To begin with, he considered General Choisy's position to be perfectly capable of withholding any attack until Washington, with the main army, could cross the river and take the British in flank. And even if Cornwallis's men had fought their way through, he poses the question: where were sufficient provisions to be found to sustain an army as it marched north (or south) through country inhabited by an increasing number of eager militia?[43] He makes it sound what it obviously was, a very forlorn hope.

Washington made no demands for surrender after the British failure to break out, but instead intensified the bombardment. From the enemy guns there spouted almost continuous orange tongues of flame as round after round went screaming into what was left of the Yorktown defences, filling the air with death. Cornwallis had not a single gun still firing and to continue the fight was pointless. At 10am on 17 October – four years to the day since Burgoyne had surrendered at Saratoga – he sent a note to Washington under a flag of truce requesting a cessation of hostilities, to which Washington agreed. The next day, commissioners were appointed to work out terms, which were to be unconditional surrender. On the 19th, the British army left their entrenchments and marched out to lay down their arms in the appointed place. General O'Hara deputised for Cornwallis, who was said to be indisposed, and so Washington delegated General Lincoln to accept the surrender.

Considering the amount of metal the allied gunners had hurled at the defences, British and Hessian casualties during the siege were fairly light; only 153 men had been killed, but over double that figure (326) were wounded and somehow 70 soldiers were missing. The Americans had 23 men killed and 65 wounded, and the French 52 killed and 134 wounded. The numbers that surrendered are given differently in almost every account, and perhaps Colonel Carrington's, taken from the original muster rolls, is the most exact. A total of 7,073 officers and men laid down their arms, and 144 cannon were handed over. In addition 30 transports, 15 galleys and a frigate with about 900 seamen were surrendered to the French.[44]

It was indeed a clamant disaster and, until recent times, the largest and most lamentable surrender a British army had endured. It was also virtually the end of the war. Ironically, on the same day that Cornwallis asked for terms, Clinton at last set sail with 7,000 troops aboard a convoy of 25 ships of the line, two fifties and eight frigates.[45] On 24 October, a schooner met the fleet with news of the surrender and Clinton sailed back to New York. It is entirely possible that had he been able to muster this force in time to reach the Chesapeake on 11 October, the day he had given Cornwallis, the outcome at Yorktown, and indeed the whole future of the war, might have been very different. For he had a fleet perfectly capable of defeating de Grasse. As at Saratoga, so at Yorktown, destiny chose other paths.

* * *

When Cornwallis arrived in Virginia in May 1781, to begin what became the Yorktown campaign, British fortunes in America were at a very low ebb. Over the course of the next few months, they were to sink beyond redemption. This was due to a whole series of mostly unnecessary mistakes and a good deal of muddling, contradiction and backbiting by senior officers. Seldom were there to be found clear judgement, promptitude of decision, quickness of perception or even good common sense present in the deliberations and actions of the generals and admirals.

Undoubtedly the most important single contributory factor to the disastrous débâcle of Yorktown was the arrival of de Grasse at Chesapeake Bay from the West Indies with many more ships than might have been expected. It is certainly possible to censure Rodney for the consequences of this, because he had failed to carry out his orders to watch the Frenchman, and when, at a late hour, his faulty intelligence did indicate the position, he did not seem to interpret the information correctly, sending fewer ships in pursuit than he could have done. Furthermore, he failed to pass on the information either to Clinton, Graves or even Hood. It is less easy to be critical of Graves without greater knowledge of prevailing conditions, but Hood was convinced that he missed an excellent chance of victory by not attacking as the French were leaving Lynhaven Bay, and that Graves made a further error in not occupying the Bay after the engagements, while de Grasse was still at sea and de Barras (presumably) had not got there first.

But while the Admirals may have been guilty of mistakes at sea, there was plenty of opportunity for the generals to follow suit on land. Clinton did not have an easy time in the early stages of the campaign, for he was for some months in complete ignorance of Cornwallis's movements. He was rightly dismayed at the latter's appearance in Virginia, and the government was unable to give him the troops he wanted for his Pennsylvania pincer movement, which although not approved by Cornwallis, was probably the soundest strategic plan. It is an unfortunate fact that throughout the Carolinas and Yorktown campaigns, Clinton was not firm enough with his deputy; admittedly, Cornwallis had virtually an independent command, but he was as impetuous as Clinton was cautious and, in this early stage of his senior appointments, an order would usually have been better than a recommendation.

The matter of a base for the protection of ships was to cause endless recrimination between the two generals, not only at the time but afterwards as well. Cornwallis never really appreciated the extreme importance of the navy, and gave the impression that he considered the selection of a base a rather tiresome interference with his other duties, but this time it was an order and had to be complied with. Whether Yorktown was the best choice is arguable. Old Point Comfort had much in its favour but, for some reason never properly disclosed, was objected to by one or two naval officers, and Portsmouth had been universally ruled out. Cornwallis cannot be seriously censured for selecting Yorktown, but he can be for not putting the army to work on the defences directly it arrived there. This was to cause him much trouble.

Probably the most controversial, and subsequently debated, military option of the campaign was whether Cornwallis should have attempted to break out of the trap he had been forced into at Yorktown, before Washington's army arrived to slam the door. There was very little chance of his being able to defeat St Simon's West Indies troops before they joined Lafayette, because they had arrived with him the day after they had landed at Jamestown Island on 3 September.[46] But Cornwallis seriously thought of trying to break through the combined Franco-American force, and considered two plans. The first was a night march down the Williamsburg road with the whole army, making use of darkness and the many ravines to come upon the enemy at first light. The second plan was to sail 2,000 infantry and six or eight guns up the York River and Queen's Creek to come in at the rear of Lafayette's men at Williamsburg, while the rest of the army attacked from the front. Tarleton, with three officers and six men, went on a reconnaissance of the Williamsburg encampment and reported the first plan to be quite feasible with every promise of success. The second was soon discarded, for it needed too great a synchronisation.[47]

However, on receipt of Clinton's dispatch of 6 September, in which he said he would join Cornwallis with 4,000 men 'already embarked' as soon as the coast was clear, Cornwallis decided to stay in Yorktown, for 'I do not think myself justifiable in putting the fate of the war on so desperate an Attempt'.[48] Clinton considered the word 'desperate' to be far too strong. Apart from the risk, Cornwallis's principal reasons for staying put were that he would have to abandon his wounded, sick and large quantities of stores; that he did not know the numbers or progress of Washington's army; and that Clinton's 'promises of relief in person were uniform, without giving me the smallest particle of discretionary power, different from holding the posts that I occupied'.[49] This latter was absolute nonsense, for Clinton had never stipulated that Cornwallis was to remain in his post, and was anyway quick to point out, 'Under what power did your Lordship act, when you moved into Virginia contrary to orders . . .'.[50]

It must have been a desperately difficult decision for Cornwallis to take, for so much depended upon it, and he will always have his apologists. But the consensus of opinion is that his decision was the wrong one, and the facts that were known at the time would seem to bear this out. By the middle of September, he was in a better position to weigh the chances of a successful relief than was either Clinton or Graves, for it all depended on the navy, and only Cornwallis, of the three, knew to what extent the French commanded the Bay and that they were in sufficient strength to make a landing of Clinton's troops most uncertain. As it happened, in due course the naval odds were to become more favourable to the British, but at the time Cornwallis was not to know this. In the circumstances, it must have been wrong for him to rely implicitly on the navy to get him out of the trap. It would surely have been better to risk a battle against Lafayette, even on somewhat uneven terms, than to await being heavily outnumbered and outgunned behind hastily prepared and inadequate defences.

Cornwallis was to make one more controversial decision before he surrendered the army. Was he right to withdraw the outer line? His reasons for doing so have already been stated. As in the case of the proposed break-out, they were influenced by Clinton's assurances of relief. It could be argued that if those caused him to abandon the attack on Lafayette, it was only logical for him to withdraw troops from a vulnerable position in order to have them readily available when reinforcements arrived. But that is not so. By holding the outer line, he would have denied the enemy a valuable position for at least two or three days and, as matters turned out, that was all he wanted – assuming, and with Digby's extra ships it is a perfectly fair assumption, that Graves could have broken through the French line of battle.

So far in this sorry tale the blame has been laid chiefly upon Cornwallis, for he was destined for the principal role, but neither Clinton nor Graves can escape censure. Here there was too much caution and sophistry; undue reliance was placed on Germain's out-of-date letters concerning Rodney's ability to neutralise any French fleet sailing from the West Indies, and this was reflected in both Clinton's and Graves's lack of urgency until it was too late. Clinton was right not to follow Washington across the Hudson, for to abandon New York would be folly, but with the reinforcements he received in August, he was strong enough to have attacked him after Graves had made excuses for not partaking in the sensible plan for an attack on Rhode Island, which might have kept de Barras out of Chesapeake Bay. This was not the only occasion when a combined operation was aborted; opportunities were missed or miscarried through the failure of naval and army commanders to co-operate.

Sadly, at a later date, Clinton and Cornwallis were to engage in much tendentious argument and accusation mainly predicated on Cornwallis's interpretations of Clinton's dispatches. Such objurgations decided nothing, and were partly misdirected, because for some of what went wrong the blame lay elsewhere. To summarise the actions of the two generals in the short Virginian campaign, there are perhaps six questions to be answered. Was Cornwallis right to have marched into Virginia? And the answer must be 'no' without Clinton's consent. Should Clinton have withdrawn Cornwallis's force from the Chesapeake and concentrated it on New York? Even with the advantages of hindsight the answer to that must be 'no'. Was Clinton right in insisting on the need for a good defensive position suitable for naval use? And to that the answer is 'yes'. Was Yorktown the best base for that purpose? Purely from looking at the map the answer is 'no', but one would like to know more about the naval objections to Old Point Comfort. Did Cornwallis take up the best defensive position? 'Yes', had he not spoilt it by withdrawing the outer line. And finally could Clinton have sailed earlier, and the answer must be a definite 'yes'.

In evaluating actions of long ago, in which conditions cannot possibly be understood exactly, it is easy to impute to the leading figures mistakes and miscalculations that they should have avoided. The answers given to the questions posed above have been made from the comfort of an armchair 200 years later, and are very much open to contradiction. But what it is difficult to dispute is that, in the case of the Yorktown Campaign, there was a disastrous concatenation of avoidable errors made by senior officers of both services, and some very untimely interference by Germain that spoilt what might have been the best plan of all, a pincer movement through the Middle Colonies.

CHAPTER 14

Independence Acknowledged: Peace Proclaimed

THE OFFICIAL NEWS of Yorktown reached London on 25 November 1781, two days before the King opened Parliament. In his speech from the throne, no mention was made of the disaster. However, it was well known that he regarded it as no great calamity and that he continued to be adamant that there should be no separation between the two countries. In government circles there was, somewhat naturally, considerable disillusionment, but no one was willing to stand up to the King and persuade him that peace was vital, and could only be obtained through considerable concessions including recognition of American independence.

In the debate on the Address, North at first made it clear that he favoured a continuation of the war in order that the colonists should submit to imperial authority, but by the time there had been subsequent debates on two motions proposed by Sir James Lowther, he had had a complete change of mind. He now favoured peace, with independence if necessary. In this he had a substantial following among government supporters, but not unanimity, for there were those who wished to continue the war until the colonists could be returned to their allegiance, and these men had for their champion Lord George Germain. There were also others, with a more realistic approach, who sought an immediate strengthening of the war against the Continental powers in the hope that at the subsequent negotiating table they might salvage some vestigial authority on the American continent.[1]

News of Yorktown caught the Opposition somewhat off balance, for there were members still not ready to recognise full independence, and the few weeks left before the Christmas recess were insufficient for the development of a satisfactory strategy. That had to wait until the new year. However, such an opportunity to harass the government could not be missed by the principal speakers in both Houses. Encouraged by a strong Remonstrance, presented to the King by the Liveried Companies of London, which included the words 'Your armies are captured; the wonted superiority of your navies is annihilated; your dominions are lost',[2] men such as Fox, Barré, Burke and the burgeoning William Pitt in the Commons developed compelling and capacious reasons for changing the ministry and ending the war. Burgoyne, now on parole and back in England, sided with Opposition speakers and supported Fox in making a case for the abandonment of British bases in America which, he pointed out, could not be held without command of the sea, and whose resources were needed for other theatres.[3] This would, of course, effectively end any further operations in that country. On 12 December, the government had a majority of 41 when Lowther's motion to end 'all further attempts to reduce the revolted colonies' was defeated, and this was largely due to many of the independent members voting with the ministers. This was a situation that would change in the new year when motions of no confidence were before the House.

Meanwhile, in the dying year, while the Opposition were putting together a viable strategy, there was a considerable degree of unrest within the ministry, most of which centred on Germain who had become anathema to almost all of his colleagues. After Yorktown, he had, as was his wont, looked around for a scapegoat and settled upon Admiral Graves, which immediately clouded still further his stormy relationship with Lord Sandwich, who was on delicate ground himself. Germain found no support for his animadversions on Graves's tactics, and shortly afterwards he was in trouble with the King and North over the replacement for Clinton, who had at last been allowed to resign his command. The obvious selection was Sir Guy Carleton, but Germain continued to pursue his personal vendetta against that general, who anyway was unlikely to accept the command while Germain remained Secretary of State.

The King may have been displeased with Germain for his obstinacy over Carleton, but he welcomed him as an ally in his intransigence over independence. Both men still firmly opposed any surrender of sovereignty, and the King instructed Germain to prepare a plan for the return of the colonies to their allegiance using those troops at present in America, for it had been agreed that no new corps would be sent. The plan he produced was based on the holding of New York with 14,000 troops, a similar number being spread between the principal ports of the south, including East Florida, and it envisaged action against enemy coastal ports, the taking of the lower Delaware counties and the repossessing of Rhode Island.[4] It is not too difficult to formulate such a plan, but it is quite a different matter when it comes to explaining the way to securing its success. His colleagues were not impressed and continued to undermine his position in the government. Indeed the Lord Advocate for Scotland, Henry Dundas, who believed the war to be lost, made it quite clear to North that he could no longer serve with Germain[5]. The First Minister was in a most unenviable position, for the King had again refused him permission to resign, and it was very clear that if his ministry was to survive far into 1783, certainly Germain, and probably Sandwich, would have to go.

While the mood of politicians and people in England complemented the winter winds that whistled mournfully over house and street, the outlook was very different in America. Washington distributed his victorious army around New York, Pennsylvania and the south, and Congress prepared as best it could for the anticipated peace negotiations. Some members were already regretting that it had been agreed in June 1781 that the French should have full control of the negotiations which, had they known it, was a slightly ironical concession, for at that time French ministers were secretly considering abandoning America and making peace with Britain. Fortunately, though, Yorktown saved Congress from this and other indignities.[6] Now it was a matter of carefully instructing Benjamin Franklin in Paris, and the peace commissioners preparing to go there, on such vital questions as boundaries, fishing rights, Loyalists and the Spanish presence.

However, these were preparations – albeit important ones – for the future and, in the meantime, Clinton in America was not thinking of peace, and Germain in London, contrary to general Cabinet thinking, was sending him instructions for limited hostilities. There were some 30,000 troops in America,[7] 12,000 of whom were in New York which Clinton considered sufficient for the security of that place, and General Leslie had 4,576 in Charleston. Clinton wrote in his narrative, 'I was, however, not in the least apprehensive for Charleston, and not much so for Savannah, as I was persuaded Greene would never attempt either . . .'. In fact he was confident enough to consider sending 2,000 troops to Jamaica,[8] but all this was before he had received the instructions from Germain.

In London the Cabinet was uncertain and divided over future American policy. Only on the determination to send no more troops, other than replacements for exisiting units, was there unanimity so far as the land war in America was concerned. It was recognised that fighting remained to be done in the Caribbean, where the policy was to reinforce the fleet at the expense of the Western Squadron in order to obtain a sufficient superiority to crush the French and Spanish squadrons and to ensure the safety of the Leeward Islands and Jamaica.[9] This was a policy strongly endorsed by Sandwich's professional advisers, but was not predicated on sound strategy, for it was in western waters that a decisive result was more likely to be achieved.

On 2 January 1782, Germain sent his dispatch to Clinton with instructions that all remaining posts and garrisons were to be retained on the Atlantic coast, and that when circumstances permitted those men could be used for combined operations with the navy against ports and coastal towns. It would seem that someone – perhaps the King – must have shown an interest in Germain's original memorandum, for some at least of his ideas were imbodied in this instruction. Chief among the reasons for retaining the various posts was that they should provide not only a refuge for the Loyalists, but bases from which they could operate. It is fascinating, but difficult, to speculate on the motives behind isolated facts such as why at this late date responsible ministers still held the view that the Loyalists might yet be capable of persuading their fellow colonists to accept some form of imperial authority.

Germain, well supported by the King, was the worst offender in this respect, and in the above mentioned dispatch to Clinton he had written, '. . . it is His

Majesty's pleasure that you do furnish them [Loyalists] with arms and ammunition, and send them such a force as you can afford to protect them in embodying themselves and disarming the rebels, and making such establishments as may enable them to resist any attempts of the Congress to subdue them.'[10] It was far too late; the Provincial Service ought to have been expanded years ago.

Parliament's Christmas recess lasted over a month, which was considered quite long enough at such a time. When it reassembled on 21 January 1782, the Opposition was ready and swung straight into action. Their ultimate goal was the downfall of North's ministry, but initially Fox, who perhaps more than any member could feel the pulse of the nation and speak for the people, concentrated his attack on Germain and Sandwich. His was by no means a lone voice, for he was ably supported by men from all the party-political groups. It so happened that, just before the recess, an event at sea provided the opponents of Sandwich with a splendid opportunity to discredit the Admiralty. Information had been received as far back as October 1781 that the Comte de Guichen was fitting out a fleet to convoy troops and stores to the West and East Indies. Two months later he sailed with 19 ships of the line, five of them carrying 110 guns apiece, and with 100 transports holding 12,000 troops destined for an attack on Jamaica, and a vast quantity of stores.[11]

To intercept this fleet, Rear Admiral Kempenfelt, a highly skilled officer, had put to sea with 12 ships of the line, a 50-gun ship and four frigates.[12] On 12 December, he sighted the French squadron some 50 miles off Ushant, and putting on all sail he was able to cut off some of their transports that had been slightly dispersed by the night's gale-force wind. He managed to capture 15 and sink two or three others,[13] but he decided the odds were too great to risk an engagement with de Guichen's more heavily armed ships. This resulted in a letter from the King to Sandwich complaining that every English admiral seemed to demand equality in numbers before engaging the enemy, which was an unfair comment both generally and in this particular case, for the action had been skilfully conducted and the result, helped admittedly by a further gale, was entirely satisfactory, for only two ships and a few transports made it to the West Indies.

But it gave the Opposition a heavy stick with which to bludgeon Sandwich, whom they had constantly censured for naval failures. Fox called the Kempenfelt expedition 'ignominious and disgraceful'.[14] The gravamen against Sandwich was that the Admirality had had plenty of notice of de Guichen's intention, and therefore more ships should have been made available for Kempenfelt. As so often, there is a divergence of opinion among historians. For example, Trevelyan, a recognised authority on the war, says 'a much stronger force just then was at the disposal of the Admiralty',[15] while Stedman, in a virtually contemporaneous account, says 'being all that were then in readiness for sea'.[16] Sandwich, in a note prepared for the Cabinet for the subsequent inquiry headed 'State of the Fleet at Home', lists the various naval commitments at that time and shows there were only 12 ships to spare. However, the list does show six ships 'wanted for the E. Indies in a month', eight for the West Indies, and seven 'Secret Service' which apparently were meant for Minorca.[17] One wonders why some of these could not have been put temporarily at Kempenfelt's disposal.

It was not only from the Opposition that Sandwich faced accusations of neglect and incompetence, for a delator perhaps even more dangerous than Fox came from his own side in the person of the Lord Advocate. Dundas was an able politician, the leader of a political faction, and a strong character who held decided views which he was prepared to press if necessary. Hitherto he had been a firm supporter of North's ministry and no friend to the Americans, but he now recognised that the war was lost, and he was determined to remove Sandwich and Germain, particularly the latter, whom he saw as a threat to peace negotiations. During the recess, he constantly pressed his case against the two ministers with North, who found himself in a difficult position, for until the King was prepared to abandon the war, North did not wish to part with Germain, nor could a successor be easily found. However, the first task was to defend Sandwich, for the debate on 20 December had been but a prelude to the main attack which Fox launched after the recess, when he moved that the House go into committee 'to inquire into the causes of the want of success of His Majesty's naval forces during this war, and more particularly in the year 1781'.[18]

Lord Sandwich had been First Lord of the Admiralty three times and, after North, he was the longest serving member of the Cabinet. His performance in office should really have been proof against any Opposition attack, but by the end of the war he seemed to have an enemy in almost every political faction. He had prepared a sound defence for his management of the navy during his time at the Admiralty, which he thought would be Fox's principal arraignment in the debate that eventually opened on 7 February. Here he was on sound ground, for his achievements were considerable: he had been responsible for the coppering of the whole fleet, and this practice of sheathing a ship's bottom with copper plates not only protected it from worm but considerably cut down time spent in dock, and increased that of speed over an uncoppered ship. In addition he had improved the dockyards, and ensured that they were stacked with timber so that the run-down navy he had inherited could be increased in numbers, and its repairs be carried out; he had augmented the number of shipwrights and increased the places where line-of-battle ships might be built. He could also produce impressive figures showing how much greater the navy was in tonnage and numbers now than at the end of the last war.[19]

But in the debate his friends were somewhat disconcerted to find that Fox, perhaps because he knew he was on unsafe ground, avoided the general state of the navy. Instead he levied his charges more specifically on the navy's failure to prevent de Grasse from reaching America and thereby ensuring French naval superiority in those waters; on the failure to prevent the loss in May 1781 of the valuable convoy sailing with the spoils from the capture of St Eustatius; on the failure later that summer to prevent the junction of the Spanish and French fleets; and, once again, the failure to provide Kempenfelt with sufficient ships.

Sandwich, of course, could not personally refute these charges in the Commons, but he had taken great pains with the help of John Robinson, the Secretary to the Treasury, to brief chosen spokesmen with his and the government's case. To the first of the three charges it could be said with some justification that the fault did not lie with the Admiralty which had sent out the necessary orders, but rather with those who for one reason or another, not always through incompetence, failed to carry them out correctly. And anyway, however well commanders performed, there would always be the risk of failure because the navy was trying to cover a wide area of the world with too few ships; it sometimes had to be a choice between, say, Yorktown or Gibraltar, the Channel or the Caribbean. Nevertheless, perhaps partly because those who spoke in defence of Sandwich were not so familiar with the subject, nor so forceful in debate as he might have been, Fox's resolution of censure was lost by only 22 votes. Two weeks later, when he put the motion again to a fuller House, the government majority fell to 19, with 217 votes for the motion against the previous 183. But in the Lords on 6 March, where Sandwich himself spoke, the Duke of Chandos's motion blaming the loss of Yorktown on insufficient naval support was fairly easily defeated by 72 votes against 37.[20]

Shortly before these momentous debates, Dundas had been persuaded to refrain from pressing further his attack on Sandwich, partly because one or two of his supporters had fallen away, but chiefly because he had got his way over Germain. Germain had always been Dundas's chief target, for he regarded his present policy as a danger not merely to the government, but to the kingdom, and he had been determined to play a pivotal part in his removal. The signs were clear that the government was in disarray, which was largely due to the King's insistence that the war must continue, and in this Germain was his principal ally. North, on the other hand, could afford to keep Germain only if the latter changed his views, for the loss of Dundas, and probably others, would be disastrous. Something of an impasse had been reached, which might have been solved if the capable Charles Jenkinson, Secretary-at-War, had accepted the American Department, for that appointment would have satisfied King and Cabinet. However, eventually the King relented, offered Lord George a peerage, which all along had been the price of his going, and found Welbore Ellis, a boring nonentity, to take his place.

Lord George Germain duly resigned at the beginning of February 1782 when the King, at Germain's special request, created him a viscount. Shortly afterwards he was introduced into the Lords, where he was greeted with a singular lack of enthusiasm; almost a record number of their lordships graced the chamber for the occasion, and it would seem many of them were incapable of forgetting Minden. Germain has received an almost universally bad press, great historians such as Fortescue and Trevelyan often had difficulty in finding words bad enough to describe his performance, and their example has been followed on both sides of the Atlantic. A careful study of Lord George's life seems to indicate that their assessment needs a little adjusting.

It is probable that he never fully recovered from Minden, and he certainly found it difficult to restore his reputation. There were many vilipenders among his contemporaries eager to traduce him even to the extent of hinting at homosexuality, for which there seems to have been no evidence. He certainly made some dreadful mistakes. Much of the blame for Saratoga can be attributed to him. During the southern campaign, his behaviour towards Clinton and his favouring of Cornwallis was inexcusable. His organising and dispatch of reinforcements to America has been frequently criticised, and some of the mud sticks, but he had problems with shipping availability, recruitment and other troop commitments. He could be impatient, inconsiderate and not beyond showing a dash of intellectual superiority, which was naturally resented by his colleagues who were hardly ever taken into his confidence. On the credit side, there were splendid mental equipment, skill in debate, a good understanding of military problems and, from first to last, a single-minded purpose to win the war. In a colourless Cabinet he was refreshingly vital. There was much to condemn in Germain's handling of his many and heavy responsibilities, but he was certainly not, as so often portrayed, the malign figure who almost single-handed lost America.

Germain's resignation left the way clear for Sir Guy Carleton to be sent to America as Clinton's successor. What must have been one of Germain's last letters in office was written to Clinton on 7 February 1782, saying that the King had agreed to his request to resign and that he was to sail for home at the first opportunity, having handed over command to Major General Robertson.[21] Clinton received this letter on 27 April. As it happened, less than a fortnight later Carleton arrived to take over. Clinton had had a very difficult command, and it might be unfair to say that he was not up to it (although his constant requests to resign could indicate this), for it covered a vast area at a time when communications were very difficult; he was consistently short of troops (soon after taking over he lost a substantial force to the West Indies); at times he was not well served by the navy whose chief, Arbuthnot, was a very difficult colleague; and latterly, Cornwallis, his second-in-command, had proved an insubordinate subordinate. Clinton left his successor an army of 31,000 all ranks, with 2,500 British and German replacements on the way to join him,[22] but of course by now all serious land fighting was over. If Germain had not been so pig-headed, Carleton might have played a leading role in the war, and it is tempting to think that matters might have gone very differently – but speculation is seldom profitable.

By the beginning of 1782, it had become obvious that North's ministry was moving, none too peacefully, towards its close. In the course of February and the first half of March, the government was challenged in six major divisions.[23]

On 7 February, as we have seen, the Opposition directed its attack mainly on Sandwich and the Admiralty, but on the 22nd and 27th on motions from General Conway that the war should no longer be pursued for the purpose of reducing the inhabitants of America by force, the scope of the debate was widened to cover the government's whole conduct of the war. And then during March the Opposition moved in for the *coup de grâce*.

On the 22nd, Conway, after deriding the ministry's handling of the military situation in the past, asked what was their policy for the future? Welbore Ellis made an unimpressive speech which seemed to show that he at least held much the same views as those that the Opposition had held for years, in that he never believed that America should be reduced by force, but that by strongly supporting the Loyalists we should defeat the war faction and return the colonists to their allegiance.[24] This may have reflected the government's thinking, but if so it was incredibly naive. However, the Secretary-at-War in his speech returned to the strategy proposed by Germain that Britain should keep those posts she held, with the possibility of increasing their size, so that the opportunity could be taken to attack, if this could be done with a reasonable chance of success.[25] Not surprisingly, this sort of talk failed to satisfy the Opposition, or many of the independents, and the government survived by only one vote.

Five days later, when Conway again moved the resolution, similar in substance but slightly altered in form, the government was defeated by 19 votes. The only true significance to be gleaned from this vote is that there were increasing numbers of members who had either abstained or voted with the government on the 22nd, who were now showing lack of confidence in the ministry, for on the vital question of peace or war there was still considerable confusion among all factions. Conway had made it clear in his speech that the principal object of his motion was to ensure that the government adhered to North's pre-Christmas declarations that all attempts at conquest by force would cease immediately, for he had detected considerable wavering on the part of some ministers. He did not rule out the retention of posts or even defensive operations, and he was not pressing for immediate peace negotiations.[26] This appeared not to be understood by many members, who had based their vote on that part of Conway's speech that indicated a lack of confidence.

At 2am on the 28th, an hour after the division in Parliament, North once again submitted his resignation to the King, and George once again refused it. North had a problem, for he could not come out publicly with his feelings while the King held a diametrically opposite view and was determined to continue the struggle. He did, however, assure the House that in accordance with the vote there would be no further offensive warfare and that he would stay in office until the King instructed him to resign, or that the sense of the House, 'expressed in the clearest manner, should point out to him the propriety of withdrawing'.[27]

The first steps in this direction came on 8 March when Lord John Cavendish moved a number of resolutions, the most important of which amounted to a vote of no confidence in the ministry. In the subsequent debate, government speakers stressed the advantages of having North in charge of affairs rather than any member of the Opposition; particularly dangerous would be Fox and Burke, for Fox had radical ideas for parliamentary reform including the representation of the people, while Burke had advocated reducing the civil list. It was largely cautionary speeches like these that spread doubts on the merit of an alternative government, and gave North a majority of 10.[28] But it was a week later, on the bodeful Ides of March, that Nemesis eventually caught up with him.

After such a close vote on the 8th, the government could not fail to read the writing on the wall and attempts were made, with the King's reluctant blessing, to form some sort of coalition. But the Marquess of Rockingham, who led the Opposition, would not agree. Instead, on that 15th of March, Sir John Rouse, specially chosen as being a well-known independent, moved a vote of no confidence in the government. In the course of the debate, Opposition members ranged widely and most critically over the whole performance of North's ministry from its inception. The question of a coalition was raised again and received fair support after some preliminary misunderstanding of its precise meaning had been corrected, but it had little chance of success while Rockingham's faction continued to oppose it. Rouse's motion was lost by nine votes, but there was now no one, with the important exception of the King, who thought that the ministry could possibly continue. As it turned out, this was to be the last division the government had to face on a vote of no confidence.

Matters moved fairly swiftly in the next few days. The Opposition let it be known that on the 20th they would bring a motion for the removal of the King's ministers from office, and North had reliable information that it would be carried. The King was still adamant that he would have no dealings with Opposition leaders, and it was at this time that he drew up a memorandum indicating his intention to abdicate.[29] But North worked on him and by the morning of the 20th, had worn down his stubbornness, and removed what thoughts he had on abdication.

It is perhaps appropriate that a ministry which in the course of 12 years had had to ride so many storms should go out, if not in a storm, at least in a squall. That afternoon, when Lord Surrey rose in the House to propose the motion, North too was on his feet and an uproar ensued as to who should give way. The Speaker favoured North, but on his proposal for adjournment there was further outcry, followed by a short stormy debate on a motion by Fox that Surrey should speak first, until Surrey eventually gave way. North then declared that he opposed the motion for the simple reason that he had already tendered his resignation to the King, which had been accepted, and that therefore his ministry was dissolved.

Much has been said that is not complimentary to Lord North; both British and American historians have treated him harshly, but as a man there was much more good than bad in him. For 12 years he had presided over a ministry with disastrous consequences, but the many mistakes, muddles and mismanagements were not all of his doing, for his team mostly lacked talent and were frequently fractious. His misfortune was that the gods had given him charm and courage, but had withheld the strong and ruthless personality that the times demanded. As a consequence, he was guilty of too great a delegation, which allowed ministers to run their own affairs in isolation, for he lacked the strength, or indeed the energy, to dominate the Cabinet and give firm and inspiring leadership. He had served a difficult King with exemplary patience and loyalty, and at the time he was about the only man capable of doing so. It is much to North's credit that his principal opponents, Fox and Burke, were prepared to pay tribute to his skills.[30]

North's ministry had at least given the country a stable government for a good many years; stability was badly needed at the time of its inception in 1770, and was just as badly needed now. The new government had men of ability who, it was hoped, might be capable of working together, for there were issues of vital importance to be dealt with, such as the problem of Ireland, the armed forces and, above all, of peace with America. In this connection, Rockingham, whom the King had reluctantly accepted as First Lord of the Treasury in preference to his own choice of Lord Shelburne, took office only on the understanding that there should be 'no veto to the independence of America',[31] a condition the King agreed to, though unhappily. In composing his new Cabinet, Rockingham retained Lord Thurlow as Lord Chancellor, filling the other posts with men who had been prominent in one or other opposition group. The Duke of Richmond became Master of the Ordnance, Admiral Keppel First Lord of the Admiralty, Barré Treasurer of the Navy, Conway Commander-in-Chief, and Lord John Cavendish Chancellor of the Exchequer. But the richest marrow in the new government's bones was undoubtedly the two Secretaries of State, Fox and Shelburne, if only they might be persuaded to work smoothly in harness, which unfortunately was to prove impossible. Burke, surprisingly, was excluded from the Cabinet but given the lucrative post of Paymaster-General.

For the remainder of 1782, the foremost task confronting politicians of the belligerent countries was the negotiation of satisfactory peace terms. Endless meetings took place in Paris, Versailles and occasionally London and Madrid, where terms were discussed, discarded and occasionally decided upon by the plenipotentiaries of the countries concerned. On the English side, there quickly arose difficulties between the two Secretaries of State, for Shelburne, although officially in charge of Irish, Home and Colonial Affairs, naturally regarded himself as responsible for America (the actual post having been abolished after Welbore Ellis).[32] But Fox, who looked after foreign affairs, which of course included France and Spain, felt he should have America as well.

Matters between them were further complicated as a result of two English emissaries being sent to Paris, Richard Oswald by Shelburne, and Thomas

Grenville (ironically George Grenville's son) by Fox. Had Fox remained in the government, this dichotomy might have caused even greater difficulties than in fact it did, but initially the two negotiators' main problem was that they had authority only to sound out the allied representatives and report back to London. Later, after some progress towards peace had been made, Alleyne Fitzherbert, the minister in Brussels, was transferred to Paris to act as the channel of communication with France, Spain and Holland.

The chief American negotiators were Benjamin Franklin, John Adams and a New York lawyer called John Jay, while France relied principally on her foreign secretary, Comte de Vergennes, with Lafayette anxious to get himself involved if it was at all possible. For Spain Charles III's chief minister, Florida Blanca, took an active interest in the American negotiations, for Spain was very sensitive to the question of Florida and the Mississippi. Franklin played the leading role from his base in Paris; he had lived in England and knew many of those in charge of affairs, and was anxious to bring matters to a head speedily, for he knew how insecure were both Rockingham's ministry, and any that might immediately succeed it. He paid lip service to the need for holding over any American agreement until peace terms with France had been settled, and kept Vergennes reasonably well-informed on his talks with the English emissaries, but in the end he was to sign the preliminaries in advance of a French settlement.

Oswald and Grenville had an extremely difficult task in trying to thread their way amid the frequent tergiversation they encountered in Paris and London. Franklin soon had the measure of Oswald, a man slow of perception, and made it quite clear what were the conditions for peace: full independence, the withdrawal of all British troops, the territorial integrity of all 13 states as they were before the Quebec Act of 1774, freedom of fishing on the banks of Newfoundland, and the settlement of the boundary with Canada. He was not prepared to do anything towards reinstating Loyalists, for their properties had been confiscated by the relevant state legislatures and Congress had no power to repeal their laws.[33]

If Franklin's diary is to be believed, Oswald was sympathetic to Franklin's views on the Loyalists, and was in favour of ceding Canada to America, which was the subject of a note from Franklin that Oswald brought to Shelburne, and which Grenville when he learnt about it, relayed to Fox.[34] In so far as this was of importance, for neither Shelburne nor Fox had any intention of even discussing the cession of Canada, it shows how hopeless it was to have two plenipotentiaries working, quite often independently, for two ministers. It also increased Fox's distrust of Shelburne.

On 1 July, in the middle of these complex negotiations, Rockingham died and Shelburne, although an unpopular man, was his obvious successor. He was highly intelligent, possessed the royal confidence in greater measure than most of his colleagues and, in spite of his earlier opposition to independence, he was now converted to the extent of granting it on the signing of a peace treaty. This qualification did not satisfy Fox who, at the end of June, proposed in Cabinet that independence should be granted immediately. His motion was defeated and Fox, who shortly after the Canada upset had indicated his intention of leaving the government, now did so. This was unfortunate, for Fox in power could be a wise and sympathetic statesman, but in opposition he quickly reverted to being an unpredictable maverick. His place was taken by Lord Grantham, Thomas Townshend became a Secretary of State and took charge of home affairs and William Pitt, so soon to rise to greater glory, replaced Lord John Cavendish as Chancellor of the Exchequer.

Cavendish was the only other member of the Cabinet to follow Fox into the wilderness, but a number of lesser members of the Rockingham party also departed. These included Burke who, with Fox, was quick to denounce Shelburne in the strongest terms as being untrustworthy, and whose ministry 'would be fifty times worse than that of Lord North'.[35] Small wonder that Franklin felt Shelburne's days would be short, and did his best to hasten the peace negotiations. But all through the summer peace remained an irritating will-o'-the-wisp. Britain's insistence over the timing of independence seemed an unnecessary sticking point and, in September, Townshend, in charge of negotiations, informed Oswald that as a very last resort Britain would be prepared to acknowledge it at once.

Other questions, however, such as boundaries and fishing rights were not so easily resolved, and matters were made no easier by two of the American commissioners – Adams and Jay – taking a violent dislike to Vergennes. They were convinced that he was acting against their interests. In this, they were not far wrong. The French were fearful of having conjured up a new state that might prove too powerful and were inclined to favour the approach of their Spanish allies over land in the unoccupied territory west of the Alleghenies and the sole navigation of the Mississippi. Vergennes was also being obdurate over the Newfoundland fishery, where the Americans were claiming exclusive access.

However, by the autumn, all parties were becoming weary of procrastination and sophistry, and eager for peace. The Americans had grave misgivings of French designs. Both France and Spain were almost bankrupt and in America, although the British had withdrawn from the South and concentrated in New York, there was still brutal mayhem between Loyalists and Patriots in the Carolinas and Georgia, which had to stop. The Spaniards too had revised their thoughts about war when, in October, the great spectacular which they and their French allies had laid on before the Duc de Bourbon and the Comte d'Artois for the final assault on Gibraltar, ended in complete fiasco.

And so by November, after the last delays and formalities had been overcome, provisional articles of peace were drawn up. These were signed between Britain and America on 30 November 1782, and although it was stated they were not to take effect until after those with France had been signed, Vergennes was very angry, for French ministers had not been consulted, and he regarded it as a gross breach of faith, and a display of ingratitude. He had a point over the latter, for the French had by the end of 1781 advanced 20 million francs to the Americans,[36] and were about to make another large loan. Nor can it be denied that theirs had been a major contribution to victory given at considerable cost, for it opened an alleyway to their own revolution. However, Franklin, who all along had been very pro-French, used his customary tact and courtesy to smooth ruffled feathers. The articles of peace between Britain, France and Spain were signed on 20 January 1783; peace with Holland was delayed a few months, but a formal truce ended hostilities.

The principal settlements contained in the peace terms in so far as they affected the American and West Indian theatres of war were that in the West Indies Britain ceded Tobago and restored St Lucia to the French (recovering them in 1803 and 1815 respectively), and got back St Kitts, Nevis, Montserrat, Dominica and the Grenadines. In America, the Spaniards kept West Florida and the British ceded East Florida to them. The boundary settled by the 1774 Quebec Act was revised, and the Americans got a large part of what had been allotted to Canada and Nova Scotia under the act, with liberty to fish on the banks of Newfoundland and the Gulf of St Lawrence. They also got the vast land area between the Alleghenies and the Mississippi, which river formed their boundary with the Spaniards, but the English retained the right to free navigation.[37] The smaller matter of the payment of debts incurred by Americans to British merchants before the war was eventually settled quite satisfactorily, but the fate of the Loyalists was shamefully resolved.

Throughout the negotiations, an indemnity for these people had been for Britain a constant and pressing cause, carried along and eventually submerged by the tide of tragical politics. Shelburne himself had urged Oswald to take the matter up forcefully with Franklin, and in this he had the support of Vergennes, but Franklin stuck to his line that Congress had no power in this matter and, as we have seen, it seems that Oswald may have been half-hearted in his advocacy. The best that could be obtained was a promise of no further prosecutions, and a mild recommendation from Congress to the states that the confiscated properties of British subjects and American Loyalists who had not borne arms should be returned. In almost every instance, nothing happened, and these wretched people, whose only crime had been that they had supported their king, were all too often left to the mercies of a mob animated by a desire for vengeance.

The Opposition found much to complain about in the treaty terms, but their most telling assault was directed at the government's shameful abandonment of the Loyalists. Lord North, in particular, made an embittered speech.[38] All of this was a trifle unfair, for Britain was not arguing from strength, and although there were still bargaining points, such as the handing over of New York and the Canadian boundary, it would have been a case of bluff easily called, and peace negotiations (and therefore the war) would have dragged on to everyone's disadvantage. In apportioning blame for a tragedy on which all parties in England looked back in sadness and anger, the Americans must bear the greater share, for in such cases an amnesty was usually given. In the end, they were probably the losers, for thousands of these good men emigrated to Canada.

The war had not been popular in England, and neither was the peace although the latter was probably as good as could be expected in the circumstances. Inevitably, Shelburne became the butt for parliament's discontent. This was largely due to his unpopularity among colleagues, which in turn was due to his seeming arrogance and aloofness. His Cabinet, starting with Admiral Keppel at the Admiralty, began to fall away or showed signs of doing so, and it was clear that he had to resign. What at first was not so clear was who should succeed him, and that was resolved in a most extraordinary way with the almost unbelievable alliance of Lord North and Charles James Fox. Members of Parliament on all sides confessed themselves amazed that North, who had had to withstand the ruthlessness of Fox's brilliant oratory, should now collaborate in senatorial grandeur with this masterful man who, for almost seven years, had viciously persecuted him. No less amazed, and alarmed, was the King who did all that he could to break this unnatural coalition by offering the Treasury first to Pitt and then to others, none of whom would accept. From 24 February, the day on which Shelburne resigned, until 2 April, when it was agreed that the Duke of Portland should head the ministry, there was no settled government in England.

The new ministry was in power when on 3 September 1783 the Treaty of Paris was signed, officially terminating the War of American Independence. A few weeks later, the last British soldiers left New York to return to England where the government was busily engaged in the usual dangerous post-war practice of reducing the army. But the principal business of this short-lived ministry concerned the controversial East India Bill, parts of which the King considered would seriously diminish the royal influence. He therefore took the opportunity, by underhand means, of ridding himself of the ministers he so hated.

The bill was passed in the Commons, but George let it be known that those in the Lords who supported it would incur his grave displeasure. Their lordships took the hint and the bill was thrown out, but much to the King's annoyance the ministers did not thereupon resign. Indeed, they carried a resolution which amounted to a vote of censure on the sovereign. The King still had a short time left in which to enjoy personal rule, and he ordered these contumacious ministers to hand in their seals of office, and not in person but through their Under-Secretaries. The new First Lord of the Treasury was the 25-year-old William Pitt whose father, Lord Chatham, had striven so hard to avoid war with the American people.

* * *

The peace negotiations and subsequent treaties require no emphasising because, from a war that Britain should never have begun, she emerged, although diminished in empire, well treated in the deliberations of the negotiators. But it is of interest to reflect upon the principal causes of defeat after seven years fighting and the loss of 43,633 men killed, and the spending of £115,654,914.[39]

There is very little doubt that properly handled the rebellion could have been crushed. In 1775, the government dithered between concession and suppression and tended to prefer the former. When this was shown not to be working, and force had to be applied, it should have been in strength: Howe, the commanding general at the time, should have been given at least the number of men his predecessor had asked for, and the necessary instructions to inflict a decisive defeat on the rebels. Small, limited operations, interspersed with conciliatory overtures, and a feeble attempt to blockade, were no good, and showed little knowledge of the minds and hearts of the American people. The business had to be accomplished quickly, before French intervention made it impossible.

Much of the trouble stemmed from the fact that North's ministry contained very little talent, and was based upon too much inter-departmental government, while what cohesive policy there was, was flawed by two serious misappreciations – the determination of the Patriots and the insufficient strength of the Loyalists. The Chinese philosopher Lao Tse once said 'Of all the dangers, the greatest is to think lightly of the foe,' and sadly that is what too many British politicians, and indeed soldiers, did for certainly the first two years of the American war. The soldiers, at least, might have got the message after Bunker Hill, but it was not until Trenton that Howe realised that in Washington he faced a general whose strategy and tactics were as sound as any he might devise, while only after Saratoga did Burgoyne realise that American Continentals were just as brave and bonny fighters as any of his own men. Even by that stage it is doubtful if those at home appreciated these important facts.

An equally serious mistake was the reliance placed on the Loyalists. British policy towards them in the early years was strangely ambivalent; they were relied upon almost to the extent of being regarded as alone able to win the war, and yet very little effort was made to form an efficient Provincial Service (a loyalist equivalent of the Patriots' Continental Army, liable for service throughout America) until it was far too late. Initially, the bulk of fighting Loyalists consisted of those volunteers who virtually pestered the authorities to allow them to form their own units, or join up individually in the British ranks. Later, when a need for their presence was greater, recruits to the ranks were not so easily found. This was not surprising, for all too often when the British had to carry out a withdrawal, Loyalists would be leaving their family and homes to the merciless vengeance of Patriots.

Britain could not wage total war in America, and great confidence was placed in Loyalists to give the necessary impetus required if restricted warfare was to succeed, for hard-line Patriots, it was hoped, might be converted by example. But numbers were never sufficient; many who had shown active loyalty before 1775 fell away when hostilities began, and at no time were there more than 7,000 Loyalists on the muster rolls.[40] A strategy that depended upon loyalist armed support just might have succeeded if it had been fully implemented from the beginning, but the lukewarm pursuance initially of this policy soon proved it to be a grave mistake. Nevertheless, Germain persisted in it to the very end.

Mistakes of a purely military nature, both personal and general, were also fairly prevalent in the course of this long war. By the middle of the 18th century the British army had not yet perfected that superb officer caste with its self-denying discipline that was to become its backbone. The quality of leadership in battle was usually very high, but there are various contemporary accounts that clearly indicate a lack of enthusiasm for the work by officers of all ranks, and very often an unhealthy eagerness to depart individually for home. The care and welfare of the soldiery, who led a harsh and brutal existence, and whose fortitude under fire was exemplary, left a great deal to be desired. It is small wonder that, at times, men were slack in their duty, despondent and low in morale. Such a state of affairs can be traced back to the higher command, although Howe certainly, and to a lesser extent Burgoyne, are known to have taken their regimental officers to task for not paying sufficient attention to the care of the wounded and the health, food and general living conditions of their men.

None of the five principal generals on the British side, or the three admirals commanding in American waters, have received much acclaim from historians, and their mistakes have been readily exposed in this volume. But the majority of these mistakes originated in London. For most of the time the chiefs were kept lamentably short of troops and ships. On at least one occasion, a substantial reinforcement was sent without regard to timing so that the soldiers arrived at the worst season of the year in a climate that quickly decimated them. Howe was never entirely sure as to what extent concession took priority over conquest, and it was largely through Germain that Howe was allowed to leave Burgoyne unsupported. Then there was the question of distance and subsequent confusion, for instructions from London took a long time to come and were often contradictory.

Whatever may be said to the contrary, there was nothing much wrong with the actual generalship in the field. Certainly the occasional strategical (and tactical) mistake was made that did not originate in London, but a number of battles were well won and the administration was as good as might be expected in a very difficult country amid a predominantly hostile population and with a supply line stretching back more than 3,000 miles.

At times there was friction between Clinton and Arbuthnot, but the only serious disruption in the command structure belongs to Clinton and Cornwallis. Their very imperfect relationship was undoubtedly damaging to the

conduct of British operations, although it could not be said to have greatly
altered the course of the war. Clinton was right to give Cornwallis an
independent command, for in those days the distance from New York to the
Carolinas was too great for a commander to carry on military operations, but
'independent command' did not mean that Cornwallis could distance himself
from his commander-in-chief completely, and disregard his instructions. Here
again the fault lay to a large extent in London, where Germain's correspondence with Cornwallis was an encouragement to the latter to consider himself
the equal of, and entirely independent of, Clinton. Writing of the closing days
of the Carolina campaign, Clinton makes out Cornwallis to be a bad general.[41]
In fact, he was a good and popular general, although he did make one or two
bad mistakes in that campaign.

There is one further debatable military point. Was it a mistake, as is often
averred, to employ Indians to fight? It angered the Americans, but then so did
the use of German mercenaries. Both were very necessary; the Germans
because the British Army simply did not have enought soldiers to carry out its
commitments, and the Indians because the Germans and the British were often
fighting the wrong sort of war. This was also the reason for another
controversial decision, to skim the cream of the infantry regiments to form
light companies. Certainly the Indians, who at this time had yet to emerge from
the cocoon of a primitive way of life, caused a large amount of trouble; but
their value as the eyes of the army was considerable, for accurate intelligence
was very hard to come by with so many unco-operative inhabitants, and the best
of the Indians proved invaluable scouts. Similarly, the light companies (first
adopted for each regiment in 1770),[42] with their men trained to move quickly,
shoot accurately and live hard, were essential in the type of almost jungle
warfare so often encountered.

The war for American independence need never have been fought; for
Britain it was a painful trek through a valley of humiliation paved with
mistakes, muddles and mismanagements. But it was not all loss, for defeat can
instruct and correct, and the British were to use their undoubted ability to
transmute disaster into gainful advantage. An opportunity would soon be
taken to examine the whole tapestry of our polity, and to pluck out the many
false stitches. For the continental contestants however, particularly France and
Spain, the war was more damaging. Their hopes of territorial expansion had
mostly perished unfulfilled. While Spain's commercial empire was only
indirectly threatened, and would survive a few years yet, France's generous
financial support given to the Colonists was partly the reason for her
forthcoming clatter into anarchy. Only the Americans emerged triumphant. A
new nation had been born which, founded on native genius tempered in the
fires of revolutionary struggle, would in due course alter the political axis of the
western world.

Appendix I

There were no children of General John Burgoyne's marriage to Lady Charlotte Stanley, but some time after her death he formed a lasting liaison with the talented opera singer Susan Caulfield from whom he had four children. The eldest (the author's great, great grandfather), for whom Charles James Fox stood sponsor, was christened John Fox in August 1782. He had a very distinguished career in the Royal Engineers which included a prominent role at the siege of Badajoz (See the author's *'Great Sieges of History'*, Brassey's 1992). After service in the Peninsular War and his brief visit to New Orleans recounted below, he went on to be Chief Engineer to Lord Raglan in the Crimea, and ended a field-marshal and a baronet. His son, Hugh, a naval officer, won the Victoria Cross in the Crimea, and later went down while commanding HMS *Captain*, an experimental ironclad. On that occasion he had omitted to take to sea his beautiful walnut cigar cabinet, which the author now uses and treasures.

At the end of 1814 John Fox, by then a full colonel, was ordered from the Peninsular fighting to accompany General Sir Edward Pakenham, as Commander Royal Engineers, when that general was sent to command the troops about to assault New Orleans at the end of the perfectly pointless and destructive American war of 1812–15. The American garrison of New Orleans was commanded by General Jackson (later seventh President of the United States), a very tough and rugged soldier who had been wounded at Trenton as a young officer, and who richly deserved the epithet 'Old Hickory'. He had three rather hastily prepared lines of defence that were given strength by the many bayous (intersecting streams or channels) in the vicinity. Against the first line the British had been making no progress at all, and it was decided to send a raiding party across the Mississippi to capture the 20-gun battery on the right bank, which could then be used to enfilade the American defences on the left bank. As soon as the raiding force sent the success signal, Pakenham was to lead a general assault.

The raiders had difficulties with the boats and there was no success signal, nevertheless Pakenham decided to assault. The result was a complete disaster, Pakenham was killed and his successor in command pulled the troops back. A council of war was then held in a tiny one-room shack, where the body of an officer, covered by a sheet, lay on the only table. At this council it was decided not to renew the attack, even though by now the raiders had captured the battery, and Colonel Burgoyne was sent across the river to organise the withdrawal of the raiding party. There was very considerable criticism at such pusillanimity, especially when it was later learnt that the Americans were so much shaken by the capture of the battery and the serious effect it would have when another attack was mounted, that they were considering abandoning the town.

Inevitably a scapegoat was sought, and as Burgoyne was the officer who withdrew the raiders it was generally thought that it was he who had persuaded the council against further action. He never uttered a word on the matter, and it was not until many years later that Major Frederic Stovin (by then Sir Frederic) decided it was time for him to speak up. He had been Assistant Adjutant General to the force, and was the 'body' lying under the sheet. Although rendered speechless by a wound in the throat, and pronounced dead by the medical officer, he had regained consciousness during the council and could hear every word spoken around him. His story, even then only confided to friends, soon became general knowledge, and he was able to give first hand evidence that in fact Burgoyne's was the only voice that had urged the renewal of the attack.[1]

Select Bibliography

Unpublished Sources
Public Record Office:
  CO 5/92, 93, 94, 95, 177, 253, 263. CO 42/36. PRO 30/55.

Devon Record Office:
  Private letter of John Burgoyne.

William L. Clements Library, Ann Arbor, Michigan:
  Clinton and Shelburne Papers.

Published Sources
Historical Manuscripts Commission:
  Vol III, Hastings and Rawdon Correspondence.
  Vol III, 9th Report Appx Pt III.
  Various Collections 6, Knox Papers, 1909.
  Stopford-Sackville, Vol II.

Parliamentary History of England:
  Vols XVIII, XIX, XXII and XXIII.

Domestic State Papers George III:
  The Howe Inquiry.

Secondary Works
Alden, J R, *The American Revolution, 1775–1783*, London, 1954
Anbury, Captain Thomas, *Travels Through the Interior Parts of America*, Vols I and II, London, 1791
Anderson, Troyer S, *The Command of the Howe Brothers During the American Revolution*, London, 1936
Andrews, C M, *The Colonial Background of the American Revolution*, Yale University Press, 1924
Atkinson, C T, British Forces in North America, *Journal of the Society for Army Research*, Vol XVI, 1937
Bancroft, George, *History of the United States of America*, Vols IV, V, VI, New York, 1885
Barnes, G R, and Owen, J H, ed, *The Private Papers of John, Earl of Sandwich, 1771–1782*, Vols III and IV, The Navy Records Society, London, 1936 and 1938
Belcher, Henry, *The First American Civil War, 1775–1778*, Vols I and II, London, 1911
Boyd, Julian P, *Joseph Galloway's Plans to Preserve the British Empire, 1774–1788*, Philadelphia, 1941
Burgoyne, Lieutenant General, *A State of the Expedition from Canada as Laid Before the House of Commons*, London, 1780

Carrington, H B, *Battles of the American Revolution*, New York, 1877

Christie, Ian R, *The End of North's Ministry 1780–1782*, Macmillan, 1958

Clark, J C D, *The Language of Liberty, 1660–1832*, Cambridge University Press, 1994

Curtis, E E *The Organisation of the British Army in the American Revolution*, London, 1926

De Fonblanque, E B, ed, *Political and Military Episodes Derived from the Life and Correspondence of The Rt Hon John Burgoyne*, London, 1876

Digby, William, *The British Invasion From the North with the Journal of Lt William Digby*, ed James P. Baxter, New York, 1887 (reprint 1970)

Egerton, H E, *The Causes and Character of the American Revolution*, Oxford, 1923

Ewald, Captain Johann, *Diary of the American War*, London, 1979

Fisher, S G, *The True History of the American Revolution*, London, 1903

Fiske, John, *The American Revolution*, Vols I and II, New York, 1919

Ford, W C, ed, *The Writings of George Washington*, Vols IV, V and VI, G. P. Putnam's Sons, 1890

Fortescue, J W, ed, *The Correspondence of King George the Third*, Vols III and IV, 1928

Fortescue, J W, *A History of the British Army*, Vol III, London, 1902

French, Allen, *The Siege of Boston*, Macmillan, 1911

Frothingham, Richard, *History of the Siege of Boston*, Boston, 1896

Furneaux, Rupert, *Saratoga: The Decisive Battle*, George Allen & Unwin, 1971

Galloway, Joseph, *A Candid Examination of the Mutual Claims of Great Britain and the Colonies*, Philadelphia, 1775

Gottschalk, Louis, *Lafayette and the Close of the American Revolution*, Chicago, 1942

Gruber, Ira D, *The Howe Brothers and the American Revolution*, University of North Carolina Press, 1974

Hartsock, J, America's First Experience in Toleration, *History Today*, Vol 43, 1993

Hatch, L C, *The Administration of the American Revolutionary Army*, London, 1904

Headley, J T, *Washington and His Generals*, Vols I and II, New York, 1847

Hibbert, Christopher, *Redcoats and Rebels*, Grafton Books, 1990

Huddleston, F J, *Gentleman Johnny Burgoyne*, Indianapolis, 1927

Kirkland, T J, and Robert M Kennedy, *Historic Camden*, Vols I and II, 1905

Lamb, Roger, *An Original and Authentic Journal of Occurrences during the Late American War from its commencement to the year 1783*, London 1809

Landers, Lieutenant Colonel H L, *The Battle of Camden*, Washington, 1929

Lecky, W E H, *A History of England in the Eighteenth Century*, Vol IV, London, 1883

Lee, Henry, *Memoirs of the War in The Southern Department of the United States*, New York, 1872

Lunt, James, *John Burgoyne of Saratoga*, London, 1975

Macintyre, Donald, *Admiral Rodney*, Peter Davies, 1962

Mackesy, Piers, *The War for America, 1773–1783*, London, 1964

Martelli, George, *Jemmy Twicher: A Life of the Fourth Earl of Sandwich, 1718–1792*, Jonathan Cape, 1962

Moultrie, William, *Memoirs of the American Revolution*, Vol I, New York, 1802

Namier, Lewis, *England in the Age of the American Revolution*, London, 1963

Oldfield, Thomas, *An Entire and Complete History, Political & Personal of the Boroughs of Great Britain*, Vols I and II, 1794

Pares, Richard, *King George III and the Politicians*, Clarendon Press, Oxford, 1953

Plumb, J H, *The First Four Georges*, London, 1974

Ramsay, David, *History of the American Revolution*, Vols I and II, Dublin, 1795

Ramsay, David, *History of the Revolution of South-Carolina, from a British Province to an Independent State*, Vols I and II, Trenton, 1785

Riedesel, Baroness Fredericka von, *Letters and Journals Relating to the War of the American Revolution*, trans. William L Stone, New York, 1867 (reprint 1968)

Robson, Eric, *The American Revolution In Its Political and Military Aspects, 1763–1783*, London, 1955

Rodger, N A M, *The Insatiable Earl: A Life of John Montagu, Fourth Earl of Sanwich*, HarperCollins, 1993

Ross, Charles, ed, *Correspondence of Charles, First Marquis Cornwallis*, Vol I, John Murray, 1859

Sanger, Ernest, *Englishmen at War: A Social History in Letters 1450–1900*, Alan Sutton, 1993

Seymour, William, *Decisive Factors In Twenty Great Battles of the World*, Sidgwick & Jackson, 1988

Seymour, William, *Yours To Reason Why*, Sidgwick & Jackson, 1982

Simcoe, Lieutenant Colonel John, *A Journal of the Operations of the Queen's Rangers*, Privately printed in Exeter, 1787

Smith, P H, *Loyalists and Redcoats*, Virginia, 1964

Smith, William H, *Life and Public Services of A St Clair*. Arranged and annotated by W H Smith, Cincinnati, 1882

Spinney, David, *Rodney*, George Allen & Unwin, 1969

Stedman, C, *A History of the Origin, Progress, and Termination, of the American War*, Vols I and II, London, 1794

Stevens, B F, ed, *The Campaign in Virginia, 1781*: an exact reprint of six rare pamphlets on the Clinton-Cornwallis controversy, London, 1888

Stone, Will L, *The Campaign of Lt Gen John Burgoyne and the Expedition of Lt Col Barry St Leger*, New York, 1893, (reprint 1970)

Tarleton, Lieutenant Colonel, *A History of the Campaigns of 1780 and 1781 In The Southern Provinces of North America*, London, 1787

Trevelyan, G O, *The American Revolution*, Vols I–IV, New edition, Longmans, Green & Co, London, 1905

Trevelyan, G O, *George The Third and Charles Fox: The Concluding Part of The American Revolution*, Vols I and II, Longmans, Green & Co, 1915

Valentine, Alan, *Lord George Germain*, Clarendon Press, 1962

Van Tyne, C H, *The American Revolution*, London, 1905

Van Tyne, C H, *The Founding of the American Republic: The Causes of the War of Independence*, Vol I, New York, 1922; *The War of Independence*, Vol II, London, 1929

Van Tyne, C H, *The Loyalists in the American Revolution*, New York, 1929

Vulliamy, C E, *Royal George*, Jonathan Cape, 1937

Wallace, Willard M, *Appeal to Arms: A Militiary History of the American Revolution*, New York, 1951

Wickwire, Franklin and Mary, *Cornwallis and the War of Independence*, Faber and Faber, 1971

Wilkin, W H, *Some British Soldiers in America*, London, 1914

Willcox, William, ed, *The American Rebellion: Sir Henry Clinton's Narrative of His Campaigns, 1775–1782*, Yale University Press, 1954

Willcox, William, The British Road to Yorktown: A Study in Divided Command, *The American Historical Review*, Vol III, 1946

Wright, Esmond, ed, *Fire of Liberty*, Hamish Hamilton, 1884

Wright, J, ed, *Speeches of the Rt Hon Charles James Fox*, Vol I, London, 1815

Source Notes

Chapter 1: Colonial and Political Background
1. Hartsock, *History Today*, Vol 43, January 1993, pp22, 23
2. Van Tyne, *Causes of the War of Independence*, p57
3. Hibbert, p5
4. *Ibid.*
5. Fiske, I, p15
6. Fisher, p34
7. Egerton, p51; Andrews, pp71, 72
8. Andrews, pp101–103
9. Van Tyne, *Causes of the War of Independence*, p69; Andrews, p16
10. Van Tyne, pp125, 126
11. Robson, p225
12. Van Tyne, *Causes of the War of Independence*, pp120, 121, 145; Hibbert, p6
13. Fiske, I, p23
14. Vulliamy, pp97, 98
15. *Ibid*, p98; Fortescue, *HBA*, III, pp29, 33
16. French, p44, 45; Egerton, p95
17. French, pp62–66
18. Van Tyne, *The American Revolution*, p19; Fiske, I. p73
19. Fisher, pp105–107
20. Fiske, I, pp94–97; Alden, p7; French, pp88–90
21. Fiske, I, pp78, 79
22. Fisher, p122
23. Fiske, I, p110
24. Fortescue, *HBA*, III, p48
25. *Ibid*, pp148, 149
26. *Ibid*, p149

Chapter 2: The Shooting Starts
1. Rodger, p233
2. Atkinson, *Journal of the Society for Army Research*, XVI, 1937, p4; Curtis, pp2, 4
3. Curtis, pp10, 11
4. Fortescue, *HBA*, III, p170
5. Fortescue, *George III Correspondence*, III, p276
6. Ewald, pXIX
7. Atkinson, pp5–23
8. Hibbert, p335
9. Van Tyne, *The Loyalists in the American Revolution*, p182
10. Anderson, p314

Chapter 2: *continued*
11. Fiske, I, p176
12. *Ibid*, p147
13. Carrington, p653
14. Hibbert, p28
15. Atkinson, *Journal of the Society for Army Research*, XIX, 1940, Pt II, p163
16. Wright, *Fire of Liberty*, p28; Fortescue, *HBA*, III, p152
17. Burgoyne MSS, Devon Record Office
18. *Ibid*.
19. Hibbert, p50
20. Burgoyne MSS
21. Willcox, *The American Rebellion*, p19
22. Carrington, p110; Bancroft, IV, pp228, 229
23. CO5/92, Pt 2, pp381–385; Hibbert, p54; Fortescue, *HBA*, III, p159; Carrington, p111
24. Burgoyne MSS
25. Fortescue, *HBA*, III, p160
26. Burgoyne MSS

Chapter 3: Initiative and Inaction
1. Bancroft, IV, p292; Fortescue, *HBA*, III, p153; Alden, p45; Hibbert, p89
2. Alden, p47
3. Fiske, I, pp165, 166
4. Fortescue, *HBA*, III, p162
5. Alden, pp52, 53; Bancroft, IV, pp297–299
6. Trevelyan, II, p280; Carrington, pp121–124
7. Hibbert, pp97, 98
8. *Ibid*. p99
9. Historical Manuscripts 9th Report, Pt III, Appx, pp85b, 86a
10. *Ibid*, p81
11. Lunt, p86
12. Willcox, *The American Rebellion*, pp11, 12 (Note 3)
13. Wright, *The Fire of Liberty*, p55
14. Fortescue, *HBA*, III, p173
15. Frothingham, p279. This acknowledged expert on the siege of Boston was never able to discover actual numbers of Loyalists who fought on the British side.
16. Fortescue, *HBA*, III, p167
17. Fiske, I, pp159, 160
18. Parliamentary History, XVIII, p978–980
19. *Ibid*, pp992 et seq.
20. Bancroft, IV, p326; Carrington, Ch XXIII
21. CO5/93, p175; Trevelyan, I, p368; Bancroft, IV, p327
22. Fiske, I, p171; Carrington, p153
23. CO5/93, p176
24. Stedman, I, p167; Fiske, I, p172; Carrington, p154
25. CO5/93, p177
26. CO5/93, p179

Chapter 4: Missed Opportunities
1. Fortescue, *HBA*, III, pp173, 174
2. Willcox, *The American Rebellion*, pp29, 30
3. Fortescue, *HBA*, III, p180; Wickwire, pp81, 82

Chapter 4: *continued*
4. Moultrie, I, p141
5. Carrington, p189; Bancroft, IV, pp408, 409
6. Valentine, p134
7. Belcher, II, p150
8. Howe, Narrative, p45
9. Fortescue, *HBA*, III, pp184, 185; Bancroft, V, pp32, 33
10. Howe, Narrative, pp4, 5
11. *Ibid*, p19
12. Anderson, pp160, 161; W L Clements Library, Shelburne Papers, Vol 88: 22
13. Fiske, I, p218
14. Ewald, p17; Hibbert, p131; Fortescue, *HBA*, III, p193
15. Ewald, pp17–25
16. Cornwallis Correspondence, I, p24; CO5/94/Pt I, pp31–35, Howe to Germain 20 December 1776; Wickwire, p93
17. Ewald, pp25–27, 382 (note 84)
18. Howe, Narrative, p68
19. *Ibid*, p19
20. Bancroft, V, p93
21. CO5/94 Pt I, pp59–61
22. Howe, Narrative, p8
23. Writings, V, p150
24. Bancroft, V, p99
25. Willcox, *The American Rebellion*, p27
26. *Ibid*, p30
27. CO5/177, pp1–13, Orders and Instructions to the Howe Brothers from the King, 6 May 1776; Anderson, p151
28. Howe, Narrative, p66; Wickwire, p92

Chapter 5: Plans and Counter-Plans for 1777
1. CO5/92, p208
2. CO42/36, p7
3. State of the Expedition, ppII–IV
4. CO5/93, pp609–613
5. Fortescue, *HBA*, III, p196
6. CO5/94, pp5–7
7. CO5/94, pp41–44
8. Gruber, p176
9. CO5/94, pp215–221
10. CO5/253, p299; CO5/94, pp287–295
11. CO5/94, pp339–344
12. Fiske, I, pp302, 303
13. HMC Stopford-Sackville, 2, p62
14. CO5/94, pp299–303; HMC Stopford Sackville, 2, p65
15. HMC Var. Coll. 6 Knox, p131; Valentine, pp207, 208

Chapter 6: The Taking of Philadelphia
1. Howe Narrative, pp15, 16
2. CO5/94, p291, Howe to Germain 2 April 1777
3. Writings, V, pp444–448
4. Howe Narrative, p60
5. W L Clements Library, Clinton MSS, Vol 21:23

Chapter 6: *continued*
6. HMC, Stopford-Sackville, II, pp72, 73
7. Willcox, *The American Rebellion*, p66
8. Writings, V, pp502, 503; Valentine, p221
9. HMC, Stopford-Sackville, II, pp74, 75
10. Fortescue, *HBA*, III, p216
11. CO5/94, pp639–658
12. Wilkin, pp154, 155
13. CO5/94, p647
14. *Ibid*, p651
15. *Ibid*, p653
16. *Ibid*, p654
17. Howe Narrative, p27
18. CO5/94, p877
19. CO5/95, pp31, 32
20. Howe Narrative, p30
21. Writings, VI, pp257, 379
22. Stedman, I, pp308–312
23. *Ibid*.
24. CO5/95, pp51–53

Chapter 7: Saratoga: A Good Beginning
1. Stedman, I, pp318, 319
2. Seymour, *Yours To Reason Why*, pp89, 90; Lunt, pp136, 137; Carrington, pp304, 307
3. Lunt, p140
4. State of the Expedition, pp6, 7; Fonblanque, p233
5. Huddleston, p134
6. Lunt, p182; Carrington, p305
7. See Lydekker, *The Faithful Mohawks* for the story of this remarkable Iroquois war-chief
8. Lunt, p155; Bancroft, V, p160; Fortescue, *HBA*, III, p224, gives a figure of 3,000 but this probably includes those not fit for duty
9. HMC Var Coll 6 Knox, p133
10. Stedman, I, p324
11. State of the Expedition, pv
12. *Ibid*, p12
13. *Ibid*, pxxi
14. *Ibid*, p40
15. Journal of Lt William Digby, pp227, 228
16. Fonblanque, p270
17. Stedman, I, p331
18. State of the Expedition, pxxii; Fortescue, *HBA*, III, p227; Lunt, p192; Seymour, *Yours To Reason Why*, p101; State of the Expedition, pxxii
19. Stedman, I, p330

Chapter 8: Saratoga: A Disastrous Ending
1. State of the Expedition, pxxii; Seymour, *Decisive Events*, p106
2. Furneaux, p129
3. *Ibid*, p131; Lunt, p200
4. Seymour, *Decisive Events*, p106; Carrington, p333; Lunt, p200
5. Seymour, *Decisive Events*, p105; Lunt, p215

Chapter 8: *continued*
6. State of the Expedition, ppxxiv–xxvi
7. Valentine, p239; Huddleston, p175
8. State of the Edpedition, pxxv
9. Furneaux, p30
10. Seymour, *Decisive Events*, p105
11. Charles W. Small, United States Park Historian
12. State of the Expedition, pxix
13. CO5/94, pp720–727; Willcox, *The American Rebellion*, p74; Valentine, p246
14. Willcox, *The American Rebellion*, p74
15. State of the Expedition, pxix
16. *Ibid*
17. *Ibid*, pl
18. Lunt, p270
19. Hudleston, p212
20. Fox, Speeches, I, pp93, 94
21. Valentine, p260

Chapter 9: The French Connection
1. Valentine, p313
2. Parliamentary History, XIX, p597
3. Fox, Speeches, I, pp102–111
4. Fortescue, *Correspondence of King George III*, IV, pp30, 31
5. Fox, Speeches, I, p114
6. Stedman, II, p9
7. Fox, Speeches, I, p117; Parliamentary History, XIX, pp767–769
8. Valentine, p313
9. Bancroft, V, p248
10. Fortescue, *Correspondence of King George III*, IV, pp56–58
11. *Ibid*, pp42, 43, 54
12. CO5/95, pp69–94 (Dispatch 8 March), pp179–206 (Dispatch 21 March); HMC, Stopford-Sackville, II, pp94, 151, 152
13. Fortescue, *Correspondence of King George III*, IV, p64
14. Willcox, *The American Rebellion*, p86
15. *Ibid*, p107; Valentine, p335
16. Trevelyan, *The American Revolution*, IV, p364
17. Bancroft, V, p273
18. Willcox, *The American Rebellion*, p96
19. Fortescue, *HBA*, III, p265
20. HMC, Stopford-Sackville, II, p152
21. HMC Var Coll 6 Knox, p150
22. Valentine, p337
23. Willcox, *The American Rebellion*, pp407, 408; W L Clements Library, Clinton Letter Book II, p270
24. Sanger, pp194, 195; Lecky, pp112, 113
25. Fortescue, *HBA*, III, pp275–280; Fiske, II, pp174, 175
26. Willcox, *The American Rebellion*, pp424–431

Chapter 10: The Carolinas: 1780
1. Fortescue, *HBA*, III, p305
2. Smith, *Loyalists and Redcoats*, p74
3. Clinton Narrative, p163

Chapter 10: *continued*

4. *Ibid*, pp163, 183, 184
5. Tarleton, pp10, 11
6. Willcox, *The American Rebellion*, pp166, 167
7. Ewald, pp229, 236
8. *Ibid*, p417
9. W L Clements Library, Clinton Papers, Vol 103:3
10. *Ibid*, Vol 101:2; Wickwire, p132
11. Clinton-Cornwallis Controversy, I, p223
12. Tarleton, pp28–30; Cornwallis Correspondence, I, p45
13. Wickwire, pp135, 136
14. Tarleton, p145
15. Clinton-Cornwallis Controversy, I, p251
16. *Ibid*, p252; Tarleton, p104
17. Clinton-Cornwallis Controversy, I, pp253, 254
18. Stedman, II, p210
19. Tarleton, pp112–115
20. Cornwallis Correspondence, I, pp58, 59
21. *Ibid*.
22. *Ibid*, PRO 30/11/64, f81
23. Willcox, *The American Rebellion*, pp227, 228
24. Cornwallis Correspondence, I, p59
25. Willcox, *The American Rebellion*, p228
26. Tarleton, p177; Stedman, p229
27. Tarleton, pp179, 180, 203–205
28. Clinton-Cornwallis Controversy, I, p284
29. PRO 30/11/82, f75; Willcox, *The American Rebellion*, p231
30. Cornwallis Correspondence, I, p74; Fortescue, *HBA*, III, pp358, 359; Willcox, *The American Rebellion*, pp231, 232
31. Willcox, *The American Rebellion*, pp479, 480
32. Sandwich Papers, III, pp261, 262
33. Stopford-Sackville, II, p192
34. Clinton-Cornwallis Controversy, I, pp310, 311
35. Trevelyan, *George The Third and Charles Fox*, II, pp126, 127
36. Stopford-Sackville, II, p190
37. HMC Var Coll 6, Knox, p168

Chapter 11: Rodney in the West Indies: 1780–82

1. Spinney, pp326. 327
2. *Ibid*, p439
3. Macintyre, pp157, 158
4. This St Leger was Lieutenant Governor of St Lucia in 1781. He could have been that Barry St Leger who was with Burgoyne in 1777. There is no *General* St Leger in the Army List of 1781; Colonel Anthony St Leger was three years senior to Barry, but his regiment (86th Foot) was not in America at that time, while Barry's (34th Foot) was
5. Spinney, p374
6. Fortescue, *HBA*, III, pp351, 352
7. Sandwich Papers, IV, p215
8. Stedman, II, p428
9. Martelli, p270
10. Sandwich Papers, IV, pp243–247

Chapter 11: *continued*
11. *Ibid*, p220
12. *Ibid*
13. Spinney, p396
14. Macintyre, p197
15. Stedman, II, p434
16. There have been many good accounts of the Battle of the Saints, but two of the
     best can be found in David Spinney's *Rodney*, London 1969, and Donald
     Macintyre's *Admiral Rodney*, London 1962
17. Gottschalk, pp387–399
18. Fortescue, *HBA*, III, p349
19. *Ibid*, p408

Chapter 12: The Carolinas: 1780
1. Christie, pp157–163; Trevelyan, *George III and Charles Fox*, II, p178
2. Valentine, pp405, 406
3. Fortescue, *HBA*, III, p358
4. Tarleton, pp210, 212
5. Stedman, II, p318
6. Tarleton, p216; The figures seem suitably exaggerated to account for the
     subsequent debacle. It is most unlikely Morgan had more than 1,100 men in
     the engagement
7. Tarleton, pp215, 218
8. PRO 30/11/84, ff83, 84; Wickwire, p275
9. Willcox, *The American Rebellion*, p262
10. Fortescue, p368 is seemingly the best account of Cornwallis's numbers
11. Tarleton, p311; Cornwallis Correspondence, I, p86
12. Cornwallis Correspondence, I, pp89, 90
13. Clinton-Cornwallis Controversy, I, p371
14. *Ibid*, p397; W L Clements Library, Clinton Papers, Vol 153:13
15. Clinton-Cornwallis Controversy, I, pp395–399; Clinton Papers, Vol 153:13
16. Fox, Speeches, I, pp377–398
17. Clinton Papers, Vol 153:13; Willcox, *American Historical Review*, 1946, LII, p12;
     Willcox, *The American Rebellion*, p279
18. Clinton-Cornwallis Controversy, I, pp424, 425
19. *Ibid*, pp428, 429
20. Willcox, *The American Rebellion*, p295
21. One of the best contemporary accounts of this siege is Stedman, II, pp365–373
22. Stedman, p378, puts the number at 400, but this must be wrong, for it would be
     almost half of Stuart's entire force
23. Cornwallis Correspondence, I, pp81–83
24. Clinton-Cornwallis Controversy, I, p362 and pp363–370
25. Cornwallis Correspondence, I, p89

Chapter 13: The Last Campaign
1. Willcox, *The American Rebellion*, p454
2. Clinton-Cornwallis Controversy, II, p33
3. Carrington, pp563–568
4. Clinton-Cornwallis Controversy, I, pp430–440
5. Clinton-Cornwallis Controversy, II, pp19–22, 24, 25
6. *Ibid*, pp31–38; PRO 30/11/68, f54
7. Tarleton, p357

Chapter 13: *continued*
8. *Ibid*, p398; Clinton-Cornwallis Controversy, II, pp29, 30; PRO 30/11/68, f28
9. Clinton-Cornwallis Controversy, I, pp493–498
10. Clinton-Cornwallis Controversy, II, pp15–17; PRO 30/11/68, ff11–13
11. Clinton-Cornwallis Controversy, II, pp26–28
12. *Ibid*, p61; PRO 30/11/68, f50
13. Valentine, p420
14. Clinton-Cornwallis Controversy, II, pp127–129
15. Fortescue, *HBA*, III, p393
16. Sandwich Papers, IV, pp125, 130
17. *Ibid*, p133
18. Willcox, *American Historical Review*, 1946, LII, p22
19. *Ibid*, p21
20. Sandwich Papers, IV, p136
21. *Ibid*, p165
22. *Ibid*, pp142, 143
23. Carrington, p615
24. Clinton-Cornwallis Controversy, II, pp49–55, 62–65
25. Willcox, *The American Rebellion*, p335
26. *Ibid*, pp336, 556
27. Clinton-Cornwallis Controversy, II, pp149, 152
28. Wickwire, p372
29. Carrington, p636
30. *Ibid*
31. Wickwire, pp366, 367
32. Clinton-Cornwallis Controversy, II, p160; Cornwallis Correspondence, I, p120
33. Clinton-Cornwallis Controversy, II, p169
34. Carrington, pp636, 637
35. Ewald, p334
36. Hibbert, p327
37. Clinton-Cornwallis Controversy, II, p175
38. *Ibid*, p188
39. *Ibid*, p172
40. Cornwallis Correspondence, I, p120
41. Willcox, *American Historical Review*, 1946, LII, pp28, 29
42. Clinton-Cornwallis Controversy, II, p211; Carrington, p640
43. Ewald, pp337, 338
44. Carrington, pp642, 643
45. *Ibid*, pp640, 641
46. Carrington, p612
47. Tarleton, p366
48. Clinton-Cornwallis Controversy, II, pp156, 157
49. Clinton-Cornwallis Controversy, I, p75
50. *Ibid*

Chapter 14: Independence Acknowledged: Peace Proclaimed
1. Christie, pp273–277
2. Valentine, p439
3. Parliamentary History, XXII, pp822, 823
4. HMC Var Coll 6 Knox, p272; HMC, Stopford-Sackville, II, p216
5. Valentine, p450
6. Rodger, p291

Chapter 14: *continued*

7. Hibbert, p335
8. Willcox, *The American Rebellion*, pp354–358
9. Rodger, p294
10. CO5/263, p244; Christie, p322
11. Martelli, p244
12. Stedman, II, p418; Trevelyan, *George III and Charles Fox*, II, p394
13. Stedman, II, p418
14. Christie, p286
15. Trevelyan, *George III and Charles Fox*, II, p394
16. Stedman, II, p418
17. Martelli, p245
18. *Ibid*, p264; Fox Speches, II, pl et seq
19. Sandwich Papers, IV, pp281–301
20. *Ibid*, p271
21. Willcox, *The American Rebellion*, p595
22. Trevelyan, *George III and Charles Fox*, II, p399 (Colonel Gerald Boyle's research).
23. Christie, p299
24. Parliamentary History, XXII, p1032; Smith, *Loyalists and Redcoats*, pp166, 167
25. Parliamentary History, XXII, p1045; Trevelyan, *George III and Charles Fox*, II, p400
26. Parliamentary History, XXII, p1068
27. Christie, p340
28. *Ibid*, pp345, 346
29. Fortescue, *Correspondence of King George III*, V, p425; Christie, p364
30. Valentine, pp378, 379
31. Bancroft, V, p533
32. Clinton-Cornwallis Controversy, I, pxxvi
33. Bancroft, V, p547
34. Lecky, IV, p230
35. *Ibid*, p237; Bancroft, V, p546
36. Robson, p174
37. Lecky, pp252, 253; Fortescue, *HBA*, III, p496
38. Parliamentary History, XXIII, pp443–458
39. Stedman, p446
40. Valentine, p437
41. Willcox, *The American Rebellion*, pp270, 271
42. Atkinson, pl

INDEX

Index